RECENT ADVANCES IN ORGANOCATALYSIS

Edited by **Iyad Karamé** and **Hassan Srour**

Recent Advances in Organocatalysis
http://dx.doi.org/10.5772/61548
Edited by Iyad Karamé and Hassan Srour

Contributors

Hideto Miyabe, Ruimao Hua, Sushmita Roy, Taek Hyeon Kim, Quynh Pham Bao Nguyen, Wei-Cheng Yuan, Wen-Yong Han, Yong-Zheng Chen, Bao-Dong Cui, Xiao-Ying Xu, Luis C. Branco, Satish Awasthi, Shrawan Kumar Mangawa

Published by InTech

Janeza Trdine 9, 51000 Rijeka, Croatia

Publishing Process Manager Dajana Pemac

Technical Editor SPi Global

Cover InTech Design team

First published September, 2016

Printed in Croatia

Additional hard copies can be obtained from orders@intechopen.com

Recent Advances in Organocatalysis, Edited by Iyad Karamé and Hassan Srour

p. cm.

Print ISBN 978-953-51-2672-0
Online ISBN 978-953-51-2673-7

For EU product safety concerns:
IN TECH d.o.o., Prolaz Marije Krucifikse Kozulić 3, 51000 Rijeka, Croatia,
info@intechopen.com or visit our website at intechopen.com.

Meet the editors

Iyad Karamé, PhD is a Full-time Professor at the Faculty of Sciences in the Lebanese University in Beirut, and the Director of the "Laboratory of Catalysis Organometallic and Materials (LCOM)," Department of Chemistry. He got his PhD degree from the Claude Bernard Lyon 1 University in France in January 2004. He was an Assistant Professor and Researcher (ATER) at the Ecole Normale Supérieure de Lyon, France, for 1 year (2004–2005), an invited researcher at the Leibniz-Institut für Katalyse in Rostock, Germany (2005–2006), and then at the Laboratory of Organometallic Chemistry of Surface, CPE Lyon, France, till 2008. His principal axes of research are organometallic and green catalysis and organic synthesis for different applications (chelating macrocycles, ligands for metal complexes, CO_2 and glycerol valorization, organocatalysts).

Hassan Srour, PhD is a research and development engineer at the École supérieure de chimie physique électronique de Lyon (C2P2-CPE de Lyon). He got his PhD degree from University of Claude Bernard Lyon 1 in France in October 2013. He was a postdoctoral fellow at the Ecole normale supérieure de Lyon (ENS), France (Nov. 2013–Mar. 2016). His principal axes of research are organic synthesis, polymer electrolytes, electrochemistry, and catalysis for different applications (energy storage systems and CO_2 valorization).

Contents

Preface

Organocatalysis has recently attracted huge attention. This branch of catalysis was realized as a fundamental field that provides wide families of catalysts for important achiral and stereoselective organic transformations.

Given the scope and the diversity of accessible transformations, metal-catalyzed reactions have become major tools in organic synthesis that will undoubtedly continue to have an important impact in the future. However, this type of reactions suffer from drawbacks such as the high cost and toxicity of the transition metal catalysts employed in some cases and the problems that their residues can cause mainly in pharmaceutical products. Alternatively, over the last years, a metal-free approach known as organocatalysis has reached a level of reliability that allowed researchers to discover new catalytic systems based on engagement of new or early-prepared organic molecules as organocatalysts. Organocatalysis meets green chemistry principles, especially the reduction of toxicity and chemical accidents and the biodegradability.

Certainly, the first organocatalysis known in nature is the hydrogen bond interaction between a hydrogen atom and an electronegative atom in biological systems. Inspired from that, recently, the utility of hydrogen bond has been widely investigated, leading to the discovery of novel chiral organocatalysts for asymmetric transformations. The first chapter of this book reviews the hydrogen bond activation mode in interesting asymmetric organocatalyzed reactions. The second chapter presents a general classification of organocatalysts based on their activation mode and then extends to the recent advances in guanidine-based chiral organocatalysts. The third chapter reviews the organocatalyzed asymmetric reaction using α-isothiocyanato compounds, and the fourth chapter reports novel researches on hydrogen-bonding isothiouronium organocatalysts considering their designed concepts and synthetic applications in non-stereoselective as well as stereoselective reactions.

Carbon dioxide transformations attracted an enormous attention in the last decade. In light of this focus, the fifth chapter summarizes an overview of organocatalytic transformation of CO_2 into cyclic carbonates, 2-oxazolidinones, carboxylic derivatives, as well as the synthesis of CO_2 adducts.

Photoredox and electro-organocatalysis processes in non-asymmetric organic transformation as well as in asymmetric ones are overviewed in the sixth chapter.

Finally, green and sustainable organocatalysis is reported in the last chapter. This chapter highlights one of the twelve principles of green chemistry—the use of benign and friendlier reaction media and conditions. In this context, several approaches using water as preferential solvent, an alternative solvent such as ionic liquids, chiral ionic liquids, deep eutectic solvents, PEG, and supercritical fluids, and organic carbonates in addition to solvent-free methodologies are overviewed.

Dear readers, this volume represents a good measure of high-quality research; nevertheless, we look forward in the near future to completing this work by editing other volumes. We hope you find this book enjoyable and illuminating in similar line to our previous work "Hydrogenation – 2012." Any comments you may have are mostly welcome.

We would like to take this opportunity to thank all the contributors for their precious contributions and their cooperation in adhering to the time-tight deadlines.

We thank warmly the Lebanese University for giving us all the facilities to complete the editions.

Finally, we wish to express our gratitude to the staff at InTech for their kind assistance in bringing this book to fruition.

Iyad Karamé
Faculty of Sciences I, Lebanese University
Beirut, Lebanon

Hassan Srour
(CPE-CNRS-F)
Lyon, France

Section 1

Organocatalysis and Asymmetric Organocatalysis Through Hydrogen Bonding

Chapter 1

Hydrogen-Bonding Activation in Chiral Organocatalysts

Hideto Miyabe

Additional information is available at the end of the chapter

http://dx.doi.org/10.5772/62459

Abstract

In a recent decade, various organocatalysts have been developed to be applicable to a wide range of asymmetric reactions. This review briefly summarizes the hydrogen-bonding activation by chiral noncovalent organocatalysts. First, the differences between hydrogen-bonding catalysts and Brønsted acid catalysts are addressed. Next, the effect of hydrogen-bonding interactions on the transition states is discussed. Finally, the hydrogen-bonding activations by the typical noncovalent organocatalysts, such as thiourea, diol, phosphoric acid, Brønsted acid-assisted chiral Brønsted acid, and *N*-triflyl phoshoramide, are shown.

Keywords: hydrogen bond, organocatalyst, enantioselective, Brønsted acid, thiourea, diol, phosphoric acid

1. Introduction

The hydrogen bond is the interaction between a hydrogen atom and an electronegative atom, which plays a central role in biological systems. Recently, the utility of hydrogen bond in organic synthesis has been widely investigated, leading to the discovery of novel chiral organocatalysts for the asymmetric transformations [1–3].

In contrast to the covalent organocatalysts, such as proline derivatives, DMAP derivatives, and *N*-heterocyclic carbene (NHC) catalysts [4–10], the noncovalent organocatalysts have been mainly developed as hydrogen bond donors or proton donors. For examples, thioureas and diols are classified into noncovalent hydrogen-bonding organocatalysts. This chapter highlights the effective and unique hydrogen-bonding activation modes by noncovalent organocatalysts [11–19]. In particular, the various activation mechanisms of nucleophilic additions into C=O and C=N bonds are described.

2. Noncovalent organocatalysts

In general, noncovalent organocatalysts can be classified into hydrogen-bonding catalysts and Brønsted acid catalysts [20], although these catalysts may rely on other additional noncovalent interactions at the same time. First, the differences between hydrogen-bonding catalysts and Brønsted acid catalysts are addressed.

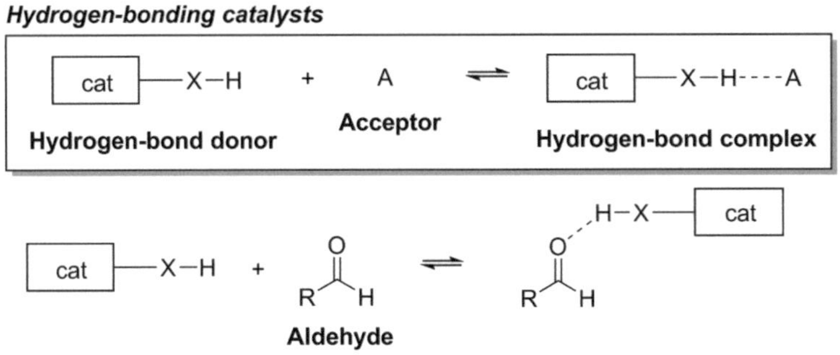

Figure 1. Hydrogen-bonding catalysts.

The hydrogen-bonding catalysts play as a hydrogen bond donor toward an electronegative hydrogen bond acceptor (**Figure 1**). The catalysts forming a hydrogen bond complex are called hydrogen-bonding catalysts. The hydrogen bonds are flexible with regard to bond length and angle. The typical bond length of a hydrogen bond is 1.5 to 2.2 Å. The hydrogen bonds are stronger than a van der Waals interaction but weaker than covalent or ionic bonds. In general, the combination of a neutral electrophile (acceptor) and a weak acid catalyst (donor) leads to hydrogen-bonding catalysis. Therefore, the nucleophilic addition to neutral carbonyl compounds, aldehyde or ketone, takes place via a hydrogen bond complex. In the case of the hydrogen bond-catalyzed reactions, a direct proton transfer from the catalyst (donor) to the electrophile (acceptor) will not occur. In other words, the hydrogen bond-catalyzed nucleophilic addition proceeds without the formation of an ion pair.

Brønsted acid catalysts play as a proton donor toward an electronegative acceptor (**Figure 2**). In general, the catalysts forming an activated ion pair are called Brønsted acid catalysts. When a catalyst (donor) is a stronger acid, the proton transfer to acceptor occurs to give an ion pair via the hydrogen bond complex. In contrast to hydrogen-bonding catalysts, the combination of basic electrophile (acceptor) and stronger acid catalyst (donor) leads to Brønsted acid-catalyzed reactions. Therefore, the nucleophilic addition to basic imine is often assumed to proceed via the formation of ion pair.

These catalysts might be simply distinguished in the point of view of proton transfer from catalysts. However, it is frequently difficult to make a clear distinction between hydrogen-bonding catalysts and Brønsted acid catalysts, because there is the equilibrium between a hydrogen bond complex and an ion pair (**Figure 2**). Moreover, Brønsted acid-catalyzed reactions can be classified into two types based on where proton transfer occurs to the substrate

http://dx.doi.org/10.5772/62459

or to the transition state. Particularly, the Brønsted acid-associated proton transfer in the transition state is closely related to the stabilization of the transition states by hydrogen-bonding catalysts.

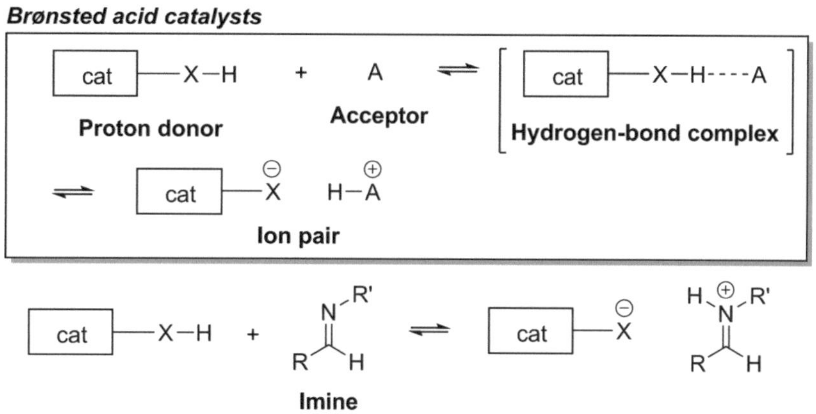

Figure 2. Brønsted acid catalysts.

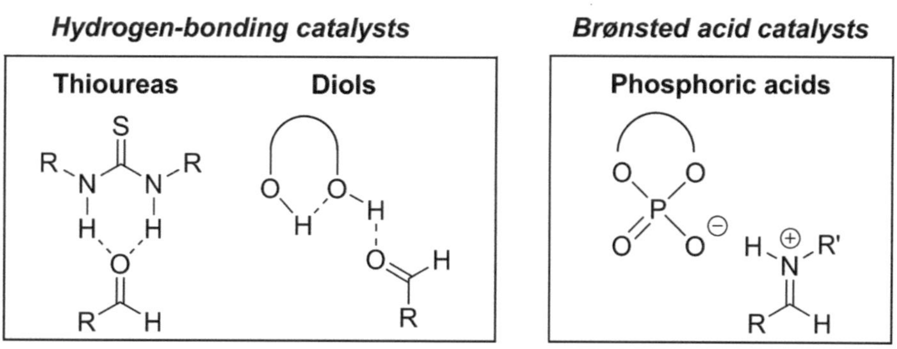

Figure 3. Activation by noncovalent organocatalysts.

Thioureas, diols, phosphoric acids, *N*-oxide, phase-transfer onium salts, etc., are well known as noncovalent organocatalysts [11–19]. Thioureas and diols are classified into hydrogen-bonding catalysts by means of the mode of activation (**Figure 3**). In contrast, phosphoric acids are generally classified into Brønsted acid catalysts.

3. Stabilization of transition states by hydrogen bond

The strength of hydrogen bond becomes larger in the charged interaction than the uncharged interaction (**Figure 4**) [21–23]. The hydrogen bond of a water molecule with a hydroxyl anion (negatively charged acceptor) is almost three times stronger than that with another water molecule (neutral acceptor) in gas phase. The hydrogen bond between a water molecule and a positively charged donor is also strong.

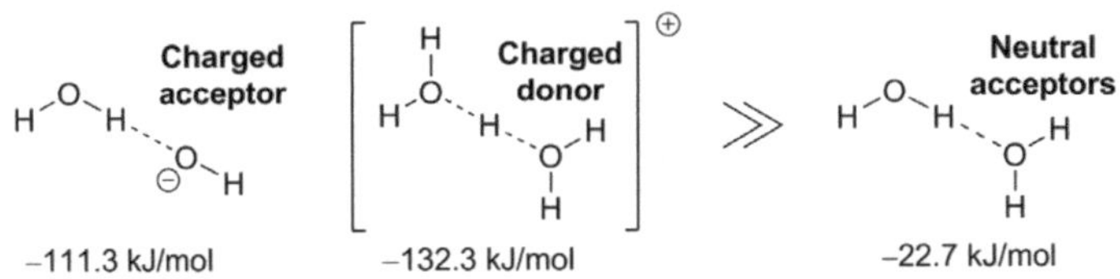

Figure 4. Strength of hydrogen bond.

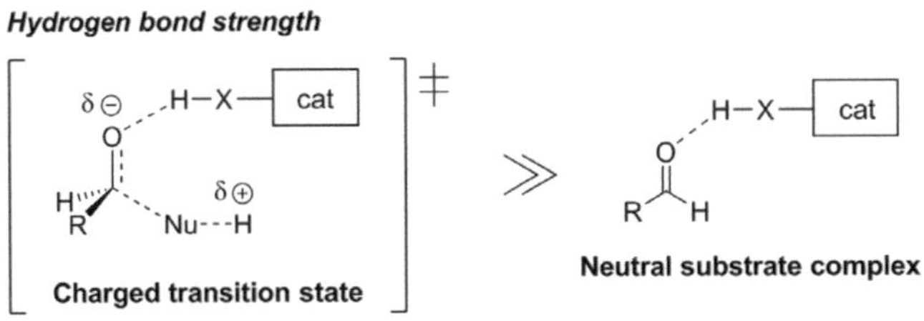

Figure 5. Hydrogen bond strength in charged transition states.

In a hydrogen bond-mediated catalysis, the functions of catalysts are both the activation of substrates and the stabilization of transition states or intermediates. Particularly, the hydrogen bonds effectively stabilize the negative charges in transition states or intermediates [24, 25], because the catalysts are bound more strongly to the charged transition states or intermediates than neutral substrates (**Figure 5**). Therefore, the study on the transition states or the charged intermediates stabilized by hydrogen-bonding interactions is of importance [26, 27], although the catalysts also affect the reaction rates by decreasing the LUMO level of neutral substrates such as carbonyl compounds and imines.

4. Hydrogen-bonding catalysts

Thioureas and diols are recognized as the typical hydrogen-bonding organocatalysts. This section highlights the hydrogen-bonding activation models and the mechanical investigations using hydrogen-bonding catalysts.

4.1. Thiourea derivatives

In 1990, the formation of crystals of diaryl ureas with carbonyl compounds as a hydrogen bond acceptor was reported by Etter's group [28]. Later, this study inspired the impressive development of thiourea catalysts. The chiral bifunctional thiourea catalyst **1** was developed by Takemoto's group (**Figure 6**) [29, 30]. Thiourea catalyst **1** catalyzed the enantioselective

http://dx.doi.org/10.5772/62459

Michael addition of 1,3-dicarbonyl compound **2** to nitroolefin **3**. The mechanism of this reaction was investigated through density functional theory (DFT) calculations by Pápai's group [31]. Between two transition states **A** and **B**, the reaction would proceed predominantly via transition state **B** due to the lower activation barrier.

Takemoto's group developed the new chiral bifunctional thiourea **5** for catalyzing the enantioselective Petasis-type reaction using organoboronic acids (**Figure 7**) [32]. In the presence of catalyst **5** and PhOCOCl, the reaction of quinoline **6** with vinyl boronic acid gave the adduct **7** in 96% ee. In this reaction, electrophilic quinoline **6** is activated as a reactive *N*-phenoxycarbonyl quinolinium salt **C**. Moreover, the chiral chelating aminoalcohol group of catalyst **5** activates the vinyl boronic acid by coordinating with the boron atom and directs the stereochemical outcome of the reaction as shown in transition state **D**.

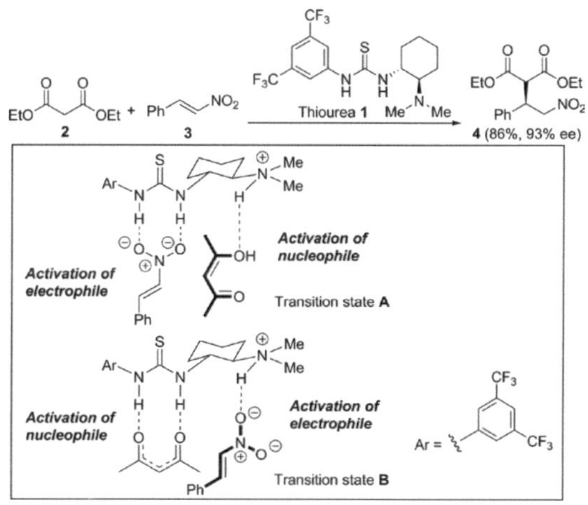

Figure 6. Thiourea-catalyzed Michael addition reaction.

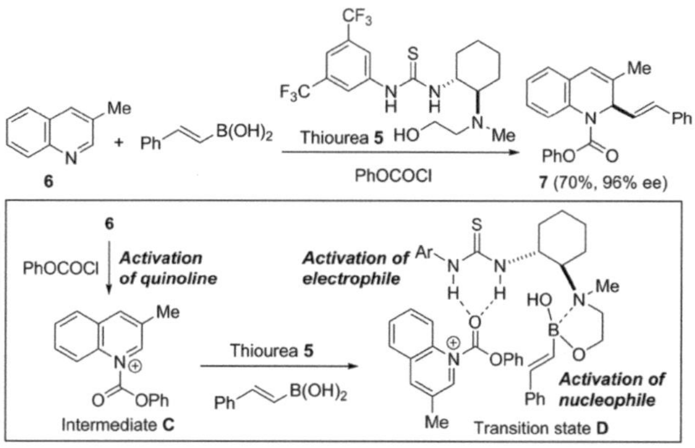

Figure 7. Thiourea-catalyzed Petasis-type reaction.

Thiourea catalyst can recognize the *in situ* generated counteranion by hydrogen bond to give the ion pair. Jacobsen's group studied the thiourea-catalyzed Pictet-Spengler-type cyclization reaction (**Figure 8**) [33, 34]. In the presence of thiourea **8**, the cyclization of indolylethyl hydroxylactam **9** gave the cyclic product **10** with good enantioselectivity. In this process, electrophile is activated as an iminium ion [35]. The thiourea catalyst **8** would promote the cyclization of **9** by abstracting a chloride on the *in situ*-generated intermediate **11**. In this proposal mechanism, thiourea **8** works as an anion receptor to form the chiral ion pair **E** involving the activated *N*-acyliminium.

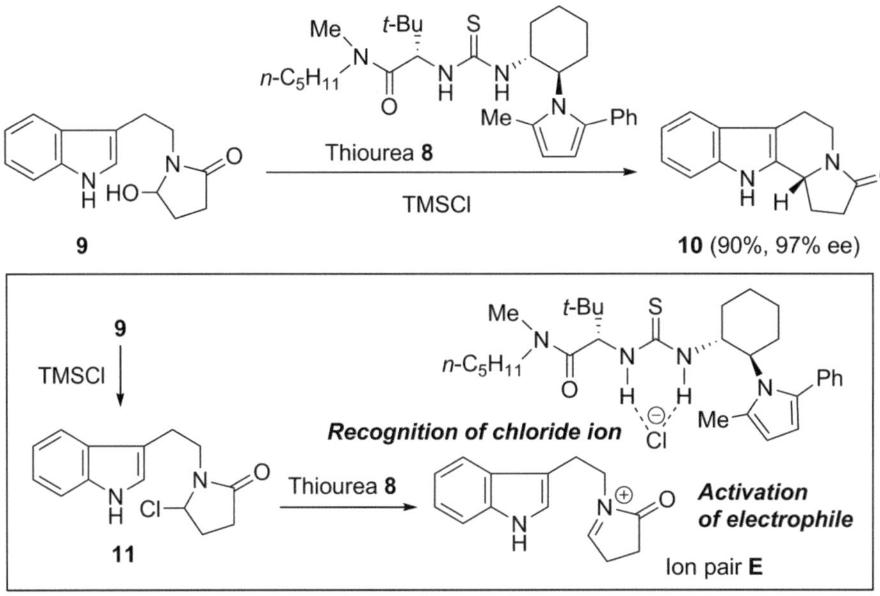

Figure 8. Thiourea-catalyzed Pictet-Spengler-type cyclization reaction.

4.2. Diol derivatives

Diols, such as *α,α,α′,α′*-tetraaryl-1,3-dioxolan-4,5-dimethanol (TADDOL), form an intramolecular hydrogen bond. (*R,R*)-1-Np-TADDOL **12** catalyzed the hetero-Diels-Alder reaction between benzaldehyde **13** and Danishefsky's diene **14** (**Figure 9**). Although the single hydrogen-bond complex **F** and the double hydrogen-bond complex **G** are the possible starting complexes, Ding's group reported that the complex **F** activated by a single hydrogen bond was supported by computational structure optimization [36].

The study on pK_a values of TADDOL analogues show that the intramolecular hydrogen bond in TADDOL analogues enhances the polarity of the second hydroxyl group and stabilizes the anion resulting from deprotonation [37]. In other words, the formation of the single hydrogen-bond complexes such as complex **F** is favored, because the increase in acidity of the second hydroxyl group on TADDOL is induced by an intramolecular hydrogen bond.

http://dx.doi.org/10.5772/62459

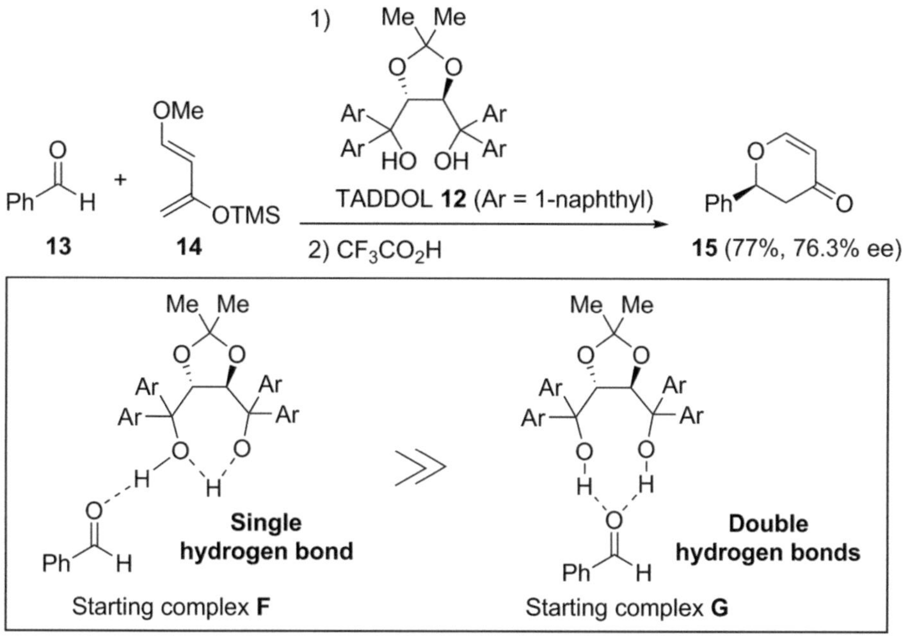

Figure 9. TADDOL-catalyzed hetero-Diels-Alder reaction.

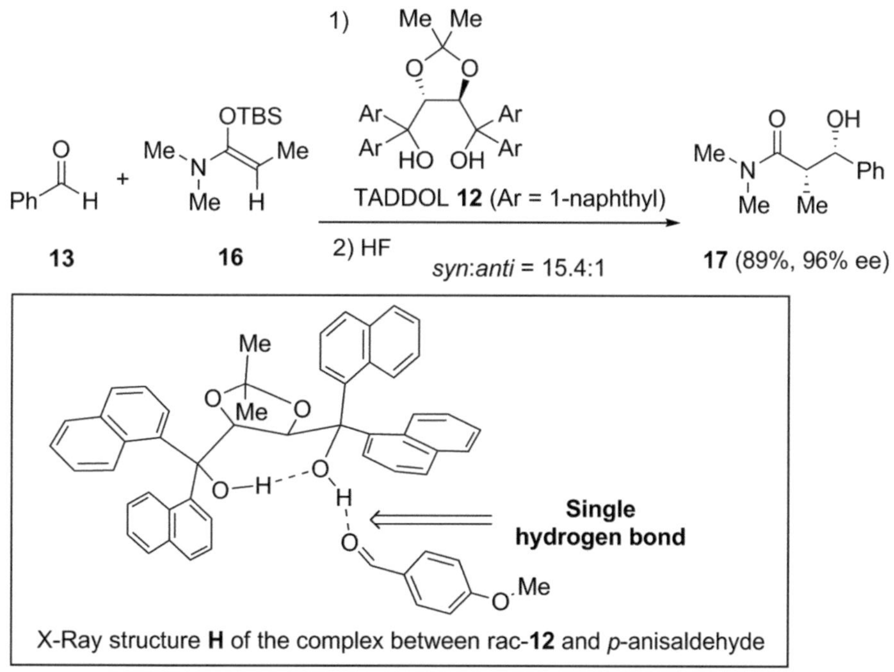

Figure 10. TADDOL-catalyzed Mukaiyama aldol reaction.

Rawal's group studied the Mukaiyama aldol reaction using TADDOL derivatives (**Figure 10**) [38]. In the presence of chiral TADDOL **12**, the reaction of benzaldehyde **13** with *O*-silyl-*N*,*O*-acetal **16** gave the product **17** in 96% ee. In this study, they succeeded the X-ray analysis of the complex **H** forming between racemic TADDOL **12** and *p*-anisaldehyde. The X-ray structure clearly supports the formation of the single hydrogen bond on the oxygen atom of *p*-anisaldehyde and the intramolecular hydrogen bond in TADDOL **12**.

5. Brønsted acid catalysts

The stronger acids are suitable for catalyzing the nucleophilic addition to basic imines as a Brønsted acid catalyst. This section highlights the activation by Brønsted acid catalyst and the investigations into Brønsted acid-associated proton transfers.

5.1. Phosphoric acid derivatives

In general, phosphoric acid derivatives are classified into Brønsted acid catalysts. The chiral BINOL-based phosphoric acid catalysts, independently developed by Akiyama's group and Terada's group, are bearing both Brønsted acidic site and Lewis basic site [39, 40]. In some cases, the bifunctional interaction of electrophilic and nucleophilic components plays a crucial role in the transition state of a rate-determining step.

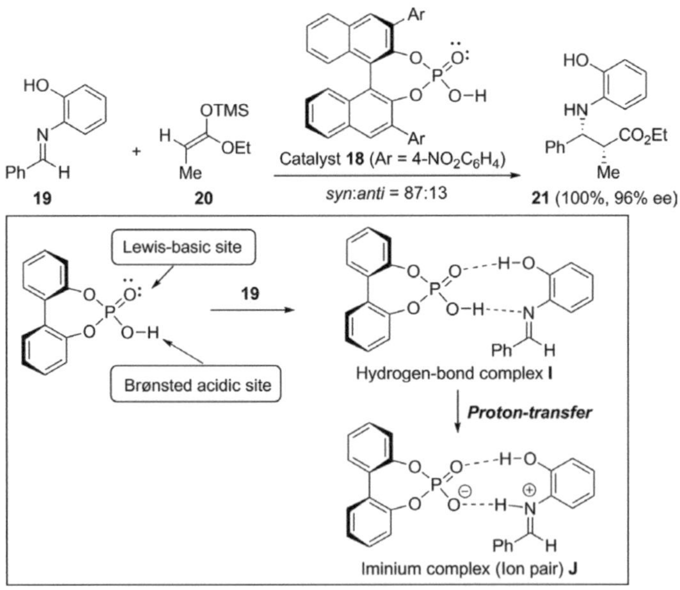

Figure 11. Phosphoric acid-catalyzed Mannich-type reaction.

Akiyama's group reported the enantioselective Mannich-type reaction using chiral phosphoric acids (**Figure 11**) [39]. In the presence of phosphoric acid **18**, the reaction of aldimine **19** with ketene silyl acetal **20** gave the adduct **21** in 96% ee. The theoretical investigation was performed using an analogous simple phosphoric acid [41]. The computational analysis supports that a double hydrogen-bonding complex **I** is the favored starting complex over a single hydrogen-bonding complex. Moreover, the calculation supports that the proton transfer from phosphoric acid to the nitrogen atom of aldimine **19** occurs to give the iminium complex **J** as a stable ion pair.

The reductions of imines with Hantzsch ester **23** have been widely investigated by the use of various chiral phosphoric acids. MacMillan's group reported that the phosphoric acid **22**–catalyzed reduction of imine, generated from ketone **24** and aniline **25**, proceeded with a good enantioselectivity [42]. In his paper, the single bonding complex between imine and Brønsted acid catalyst **22** was proposed. More recently, Goodman's group performed the DFT calculation study, which indicates that the ternary complex between imine, catalyst **22**, and Hantzsch ester **23** is important to stereochemical course [43]. The Z-imine transition state **K** is on the lowest energy route to the product (**Figure 12**).

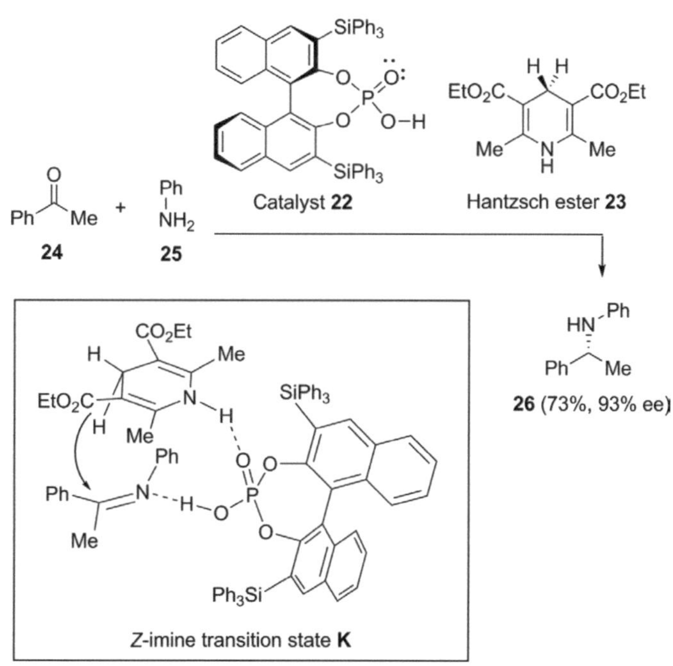

Figure 12. Phosphoric acid-catalyzed reduction using Hantzsch ester.

5.2. Brønsted acid-assisted chiral Brønsted acid

Ishihara's group developed Brønsted acid-assisted chiral Brønsted acid catalyst **27** (**Figure 13**) [44]. In the presence of catalyst **22**, Mannich-type reaction of aldimine **28** with ketene silyl acetal **29** enantioselectively proceeded to give the adduct **30** in 77% ee. The X-ray analysis of catalyst **27** indicates the acidity of bis(triflyl)methyl proton of **27** increases by intramolecular hydrogen bond between phenolic hydroxy group and oxygen atom of triflyl group, which is 2.305 Å. Therefore, the activation of aldimine **28** by the acidic bis(triflyl)methyl proton of **27** should be important.

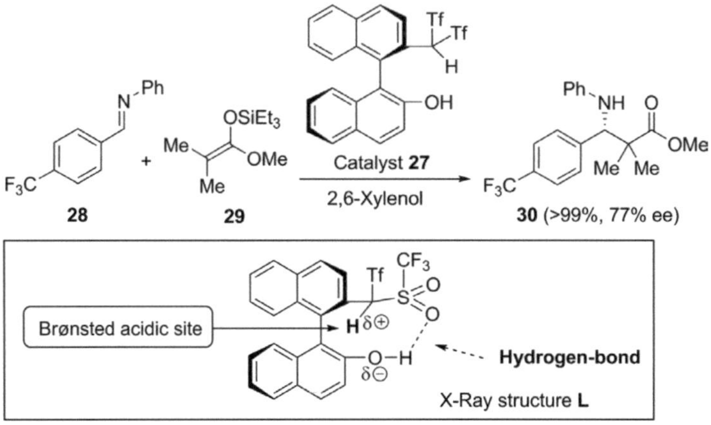

Figure 13. Catalyst **22**-catalyzed Mannich-type reaction.

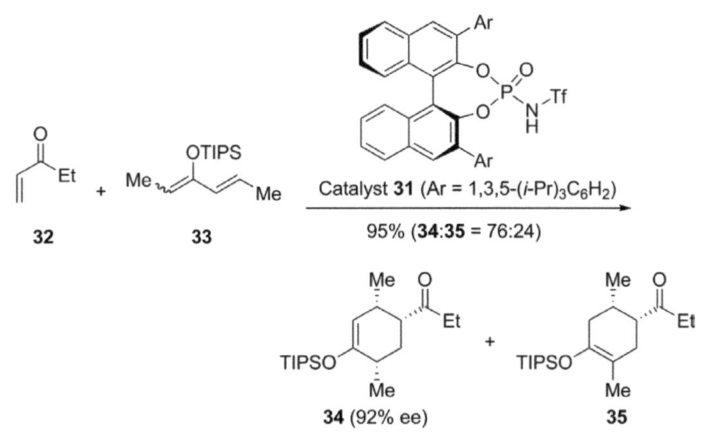

Figure 14. *N*-triflyl phoshoramide-catalyzed Diels-Alder reaction.

5.3. *N*-triflyl phoshoramide

Yamamoto's group studied the Diels-Alder reaction using chiral *N*-triflyl phoshoramides **(Figure 14)** [45]. *N*-triflyl phoshoramides are strong Brønsted acids due to a strong electron-withdrawing triflyl group. In the presence of *N*-triflyl phoshoramide **31**, the reaction of ethyl vinyl ketone **32** with diene **33** effectively proceeded to give the cyclic adduct **34** in 92% ee, accompanied by the isomeric adduct **35**. The bulky 1,3,5-(i-Pr)$_3$C$_6$H$_2$ groups on catalyst **31** are essential to achieve the reaction with good chemical efficiencies.

http://dx.doi.org/10.5772/62459

6. Concluding remarks

The utilization of organocatalysts in organic synthesis has become a subject of recent research. Particularly, chiral hydrogen-bonding catalysts and chiral Brønsted acid catalysts have developed as the highly efficient noncovalent organocatalysts for a broad spectrum of asymmetric transformations. One of the most important features of noncovalent organocatalysts is that we can use the hydrogen bond for the stabilization of transition states or intermediates as well as the activation of C=O bond in carbonyl compounds or C=N bond in imines. Because the use of organocatalysts has many advantages in organic synthesis form both economical and environmental points of view, the research into organocatalysts continues to blossom and grow.

Author details

Hideto Miyabe

Address all correspondence to: miyabe@huhs.ac.jp

School of Pharmacy, Hyogo University of Health Sciences, Minatojima, Chuo-ku, Japan

References

[1] List B, editor (2012) Asymmetric Organocatalysis 1—Workbench Edition: Lewis Base and Acid Catalysts in Science of Synthesis. Stuttgart: Thieme.

[2] Maruoka K, editor (2012) Asymmetric Organocatalysis 2—Workbench Edition: Brønsted Base and Acid Catalysts, and Additional Topics in Science of Synthesis. Stuttgart: Thieme.

[3] Dalko PI, editor (2007) Enantioselective Organocatalysis. Weinheim: Wiley-VCH.

[4] Lelais G, MacMillan DWD (2006) Modern Strategies in Organic Catalysis: The Advent and Development of Iminium Activation. Aldrichima Acta 39: 79–87.

[5] Erkkilä A, Majander I, Pihko PM (2007) Iminium catalysis. Chem. Rev. 107: 5416–5470.

[6] Mukherjee S, Yang JW, Hoffmann S, List B (2007) Asymmetric enamine catalysis. Chem. Rev. 107: 5471–5569.

[7] Gaunt MJ, Johansson CCC (2007) Recent developments in the use of catalytic asymmetric ammonium enolates in chemical synthesis. Chem. Rev. 107: 5596–5605.

[8] Wurz RP (2007) Chiral dialkylaminopyridine catalysts in asymmetric synthesis. Chem. Rev. 107: 5570–5595.

[9] Enders D, Niemeier O, Henseler A (2007) Organocatalysis by *N*-heterocyclic carbenes. Chem. Rev. 107: 5606–5655.

[10] Bugaut X, Glorius F (2012) Organocatalytic umpolung: *N*-heterocyclic carbenes and beyond. Chem. Soc. Rev. 41: 3511–3522.

[11] Sohtome Y, Nagasawa K (2012) Dynamic asymmetric organocatalysis: cooperative effects of weak interactions and conformational flexibility in asymmetric organocatalysts. Chem. Commun. 48: 7777–7789.

[12] Maruoka K (2010) Highly practical amino acid and alkaloid synthesis using designer chiral phase transfer catalysts as high-performance organocatalysts. Chem. Rec. 10: 254–259.

[13] Miyabe H, Takemoto Y (2008) Discovery and application of asymmetric reaction by multi-functional thioureas. Bull. Chem. Soc. Jpn. 81: 785–795.

[14] Hashimoto T, Maruoka K (2007) Recent development and application of chiral phase-transfer catalysts. Chem. Rev. 107: 5656–5682.

[15] Doyle AG, Jacobsen EN (2007) Small-molecule H-bond donors in asymmetric catalysis. Chem. Rev. 107: 5713–5743.

[16] Schreiner PR (2003) Metal-free organocatalysis through explicit hydrogen bonding interactions. Chem. Soc. Rev. 32: 289–296.

[17] Akiyama T (2007) Stronger Brønsted acids. Chem. Rev. 107: 5744–5758.

[18] Terada M (2010) Chiral phosphoric acids as versatile catalysts for enantioselective transformations. Synthesis 42: 1929–1982.

[19] Yu J, Shi F, Gong LZ (2011) Brønsted-acid-catalyzed asymmetric multicomponent reactions for the facile synthesis of highly enantioenriched structurally diverse nitrogenous heterocycles. Acc. Chem. Res. 44: 1156–1171.

[20] Akiyama T (2009) Hydrogen-bond catalysis or Brønsted acid catalysis? General considerations. In: Pihko PM, editor. Hydrogen Bonding in Organic Synthesis. Weinheim: Wiley-VCH. pp. 5–14.

[21] Curtiss LA, Frurip DJ, Blander M (1979) Studies of molecular association in H_2O and D_2O vapors by measurement of thermal conductivity. J. Chem. Phys. 71: 2703–2711.

[22] Xie Y, Remington RB, Schaefer HFIII (1994) The protonated water dimer: extensive theoretical studies of H_5O^+. J. Chem. Phys. 101: 4878–4884.

[23] Pudzianowsk AT (1995) MP2/6-311++G(d,p) study of ten ionic hydrogen-bonded binary systems: structures, normal modes, thermodynamics, and counterpoise energies. J. Chem. Phys. 102: 8029–8039.

[24] Jencks WP (1976) Enforced general acid-base catalysis of complex reactions and its limitations. Acc. Chem. Rev. 9: 425–432.

[25] Jencks WP (1980) When is an intermediate not an intermediate? Enforced mechanisms of general acid-base, catalyzed, carbocation, carbanion, and ligand exchange reaction. Acc. Chem. Rev. 13: 161–169.

[26] Cheong PHY, Legault CY, Um JM, Çelebi-Ölçüm N, Houk KN (2011) Quantum mechanical investigations of organocatalysis: mechanisms, reactivities, and selectivities. Chem. Rev. 111: 5042–5137.

[27] Berkessel A, Etzenbach-Effers K (2009) Computational studies of organocatalytic processes based on hydrogen bonding. In: Pihko PM, editor. Hydrogen Bonding in Organic Synthesis. Weinheim: Wiley-VCH. pp. 15-42.

[28] Etter MC, Urbañczyk-Lipkowska Z, Zia-Ebrahimi M, Panunto TW (1990) Hydrogen bond-directed cocrystallization and molecular recognition properties of diarylureas. J. Am. Chem. Soc. 112: 8415–8426.

[29] Okino T, Hoashi Y, Takemoto Y (2003) Enantioselective Michael reaction of malonates to nitroolefins catalyzed by bifunctional organocatalysts. J. Am. Chem. Soc. 125: 12672–12673.

[30] Xu X, Furukawa T, Okino T, Hoashi Y, Miyabe H, Takemoto Y (2006) Bifunctional-thiourea-catalyzed diastereo- and enantioselective Aza-Henry reaction. Chem. Eur. J. 12: 466–476.

[31] Hamza A, Schubert G, Soós T, Pápai I (2006) Theoretical studies on the bifunctionality of chiral thiourea-based organocatalysts: competing routes to C-C bond formation. J. Am. Chem. Soc. 128: 13151–13160.

[32] Yamaoka Y, Miyabe H, Takemoto Y (2007) Catalytic enantioselective petasis-type reaction of quinolines catalyzed by a newly designed thiourea catalyst. J. Am. Chem. Soc. 129: 6686-6687.

[33] Taylor MS, Jacobsen EN (2004) Highly enantioselective catalytic acyl-Pictet-Spengler reactions. J. Am. Chem. Soc. 126: 10558–10559.

[34] Raheem IT, Thiara PS, Peterson EA, Jacobsen EN (2007) Enantioselective Pictet-Spengler-type cyclizations of hydroxylactams: H-bond donor catalysis by anion binding. J. Am. Chem. Soc. 129: 13404–13405.

[35] Brière JF, Oudeyer S, Dalla V, Levacher V (2012) Recent advances in cooperative ion pairing in asymmetric organocatalysis. Chem. Soc. Rev. 41: 1696–1707.

[36] Zhang X, Du H, Wang Z, Wu YD, Ding K (2006) Experimental and theoretical studies on the hydrogen-bond-promoted enantioselective hetero-Diels-Alder reaction of Danishefsky's diene with benzaldehyde. J. Org. Chem. 71: 2862–2869.

[37] Christ P, Lindsay AG, Vormittag SS, Neudörfl JM, Berkessel A, O'Donoghue AC (2011) pK_a values of chiral Brønsted acid catalysts: phosphoric acids/amides, sulfonyl/sulfuryl imides, and perfluorinated TADDOLs (TEFDDOLs). Chem. Eur. J. 17: 8524–8528.

[38] McGilvra JD, Unni AK, Modi K, Rawal VH (2006) Highly diastereo- and enantioselective Mukaiyama aldol reactions catalyzed by hydrogen bonding. Angew. Chem. Int. Ed. 45: 6130–6133.

[39] Akiyama T, Itoh J, Yokota K, Fuchibe K (2004) Enantioselective Mannich-type reaction catalyzed by a chiral Brønsted acid. Angew. Chem. Int. Ed. 43: 1566–1568.

[40] Uraguchi D, Terada M (2004) Chiral Brønsted acid-catalyzed direct Mannich reactions via electrophilic activation. J. Am. Chem. Soc. 126: 5356–5357.

[41] Yamanaka M, Itoh J, Fuchibe K, Akiyama T (2007) Chiral Brønsted acid catalyzed enantioselective Mannich-type reaction. J. Am. Chem. Soc. 129: 6756–6764.

[42] Storer RI, Carrera DE, Ni Y, MacMillan DWC (2006) Enantioselective organocatalytic reductive amination. J. Am. Chem. Soc. 128: 84–86.

[43] Simón L, Goodman JM (2008) Theoretical study of the mechanism of Hantzsch ester hydrogenation of imines catalyzed by chiral BINOL-phosphoric acids. J. Am. Chem. Soc. 130: 8741–8747.

[44] Hasegawa A, Naganawa Y, Fushimi M, Ishihara K, Yamamoto H (2006) Design of Brønsted acid-assisted chiral Brønsted acid catalyst bearing a bis(triflyl)methyl group for a Mannich-type reaction. Org. Lett. 8: 3175–3178.

[45] Nakashima D, Yamamoto H (2006) Design of chiral *N*-triflyl phosphoramide as a strong chiral Brønsted acid and its application to asymmetric Diels-Alder reaction. J. Am. Chem. Soc. 128: 9626–9627.

Chapter 2

Isothiouronium Organocatalysts Through Hydrogen Bonding

Quynh Pham Bao Nguyen and Taek Hyeon Kim

Additional information is available at the end of the chapter

http://dx.doi.org/10.5772/62991

Abstract

The field of small-molecule organocatalysis via noncovalent interactions has attracted the attention of an increasing number of research groups from the academic as well as industrial sectors. Isothiouronium salts have been explored quite recently as a new class of hydrogen-bonding subunit for the purpose of molecular recognition of anions in supramolecular chemistry. The chemical modification of isothiouroniums is readily varied using synthetic methods to make several types of functional molecular systems. This chapter, for the first time, describes the research on hydrogen-bonding isothiouronium organocatalysts considering their designed concepts and synthetic applications in both nonstereoselective and stereoselective reactions.

Keywords: asymmetric aldol reactions, hydrogen-bonding organocatalysts, isothiouronium, reduction, reductive amination

1. Introduction

Organocatalyst has emerged as one of the hot topics in advanced organic chemistry. Although chemical transformations that use organocatalysts have been studied broadly, the field of organocatalyst was not a key area of research until the late 1990s [1]. Organocatalysts consist of small, low-molecular-weight organic compounds containing carbon, hydrogen, nitrogen, sulfur, and phosphorus and acting as catalysts without any metal. The popularity of this word is due to the notion of green chemistry as opposed to dirty organometallic chemistry. The advantages of using organocatalysts include their lack of sensitivity to moisture and oxy-

easy availability, low cost, and low toxicity, which confers a huge direct benefit in the production of pharmaceutical intermediates when compared with (transition) metal catalysts. In addition, organocatalysts are also more readily amenable than both metal-based catalysts and biocatalysts to anchor on a support with the aim of facilitating catalyst recovery and recycling [2].

The development of acid- and metal-free organocatalysts still remains a challenging task. As investigated, organocatalysts can be considered as minimal versions of metal-free enzymes, and the mechanisms and categorizations of enzyme catalysis are applied to the action of organocatalyst as well. In biological systems, hydrogen bonding plays a key role in many enzymatic reactions, both in orienting the substrate molecules and lowering barriers to reaction. Therefore, in many cases noncovalent organocatalysts depend on the formation of hydrogen-bonding adducts between the substrate and catalyst [3]. Hydrogen bonding can promote reactions by different mechanisms such as stabilizing anionic intermediates and transition states. It can also bind small anions for the formation of reactive electrophilic cations or more acidic donors can actively electrophile by protonation. Especially, for simultaneous activation of both partners, for example, nucleophile and electrophile, in a reaction, bifunctional organocatalysis has been developed so far. In this case, the close association between the organocatalyst molecule and the substrate makes the hydrogen-bonding catalysis a powerful method for inducing enantioselectivity [4].

2. Designed concepts of hydrogen-bonding isothiouronium organocatalysts

2.1. Hydrogen-bonding thiourea organocatalysts [5, 6]

Thioureas are widely recognized as highly useful templates using them powerful organocatalytic systems can be constructed. They result in a considerable acceleration of the reaction rate through hydrogen-bonding interaction. The scope of these small-molecule H-bond donors, termed thiourea organocatalysts, covers both nonstereoselective and stereoselective applications in organic synthesis. The reader may get an idea about the variety of this research field that is generally based on (thio)urea organocatalysts for *"hydrogen bonding in organic synthesis."*

2.2. Hydrogen bonding in anion recognitions: isothiouronium versus thiourea derivatives

Thioureas have also been thoroughly investigated in the field of molecular recognition [7]. Isothiouronium salts have been explored quite recently as a new class of hydrogen-bonding subunit for the purpose of molecular recognition of anions in supramolecular chemistry. Isothiouroniums were proven as prospective replacements of thioureas because such groups would enhance the acidicity of the NH moieties and therefore can function as a better binder. The examples of hydrogen-bonding isothiouronium derivatives in anion recognitions were described in **Figure 1** [8–15].

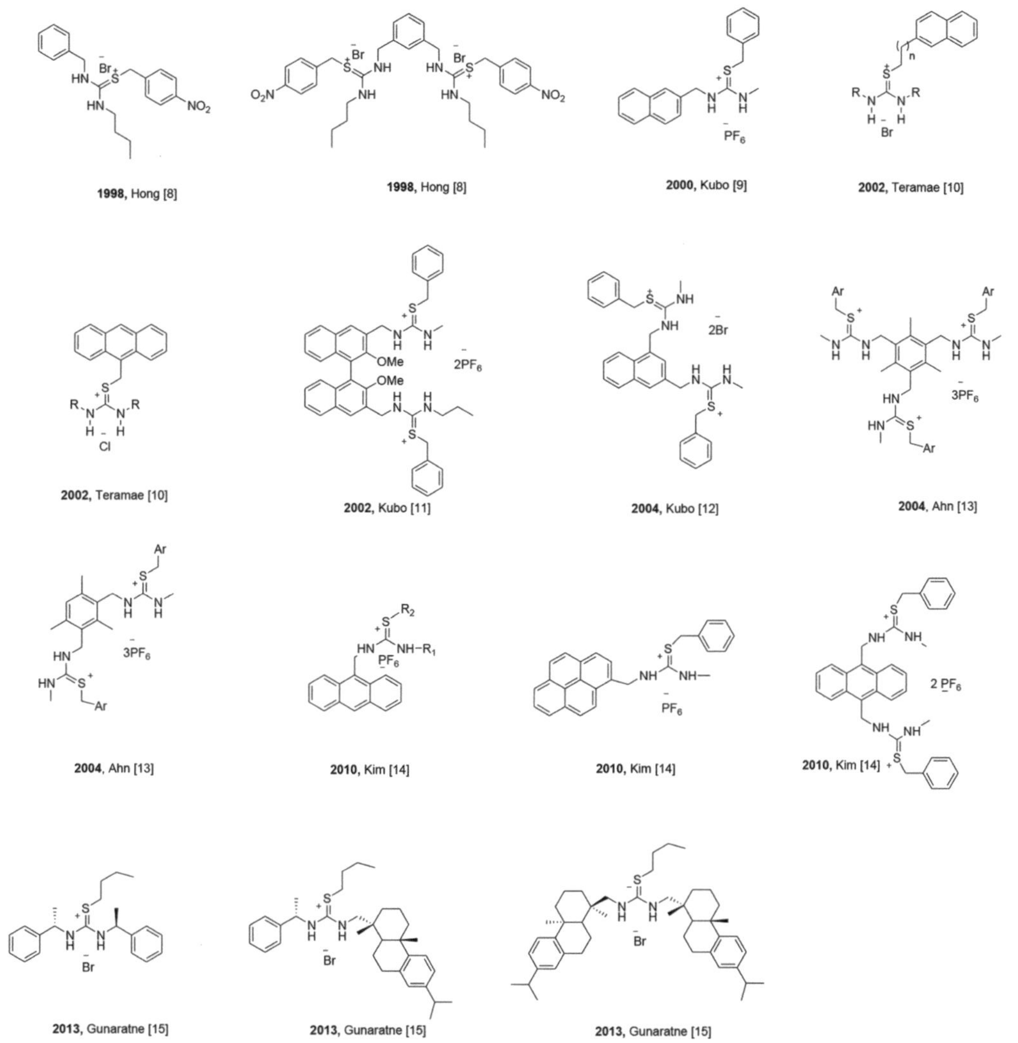

Figure 1. Isothiouronium-based anion recognitions.

2.3. Hydrogen-bonding isothiouronium organocatalysts

As previously explored, there is a close relationship and mutual interplay between molecular recognition, active site consideration in enzyme catalysis involving anions, and organocatalysis utilizing explicit hydrogen bonding. Molecular recognition is the central component of bio- and organocatalysis. Indeed, the concepts of anion bindings are the key to designing new organocatalytic transformations [16, 17]. Keeping this fact in mind, in organocatalysts, isothiouroniums can be used to improve the catalytic property of thioureas because they are better binder than anion recognitions. In addition, the chemical modification of isothiouroniums is readily varied using synthetic methods to make several types of functional molecular systems. Consequently, isothiouronium-derived catalysts have been explored as a new field in hydrogen-bonding organocatalysts (**Figure 2**) [18–23].

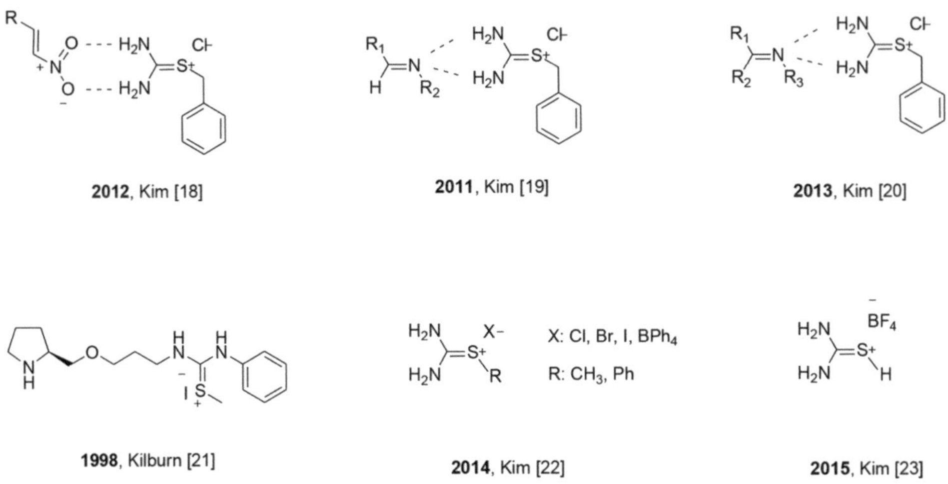

Figure 2. Hydrogen-bonding isothiouronium organocatalysts.

2.4. Synthesis of isothiouronium salts

Isothiouronium salts are typically obtained by the displacement reaction of alkyl halides with thioureas (**Scheme 1**) [24].

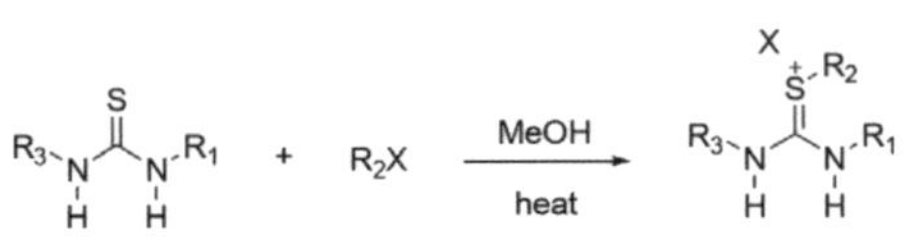

Scheme 1. Synthesis of isothiouronium salts.

It should be noted that the configurational and electronic structures of isothiouronium salts presented complexities and remained a subject for investigation (**Figure 3**) [15, 25]. The following three possibilities of isomerism have been reported ($R_1 \neq R_3$):

I. The oscillation of the carbon–nitrogen (CN) double bond (Eq. 1).

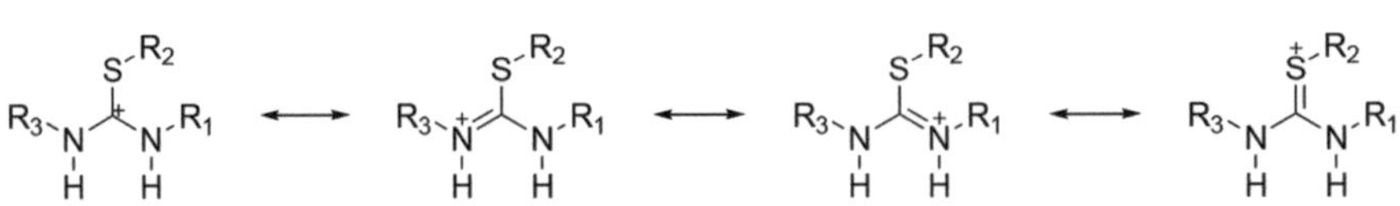

(1)

II. Hindered rotation at the CN bond (Eq. 2).

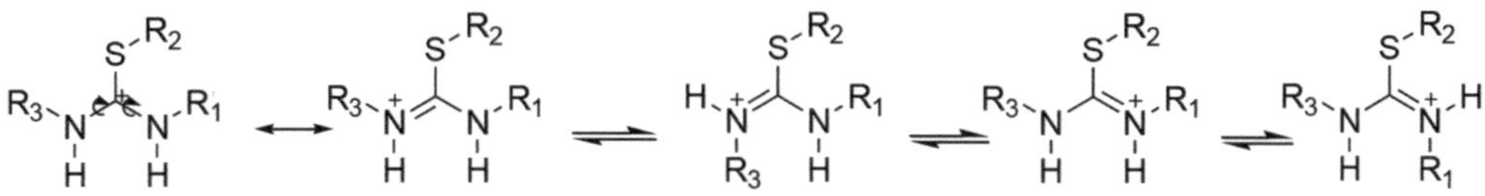

(2)

III. Syn–anti isomerism from restricted rotation at the carbon–sulfur (CS) bond (Eq. 3).

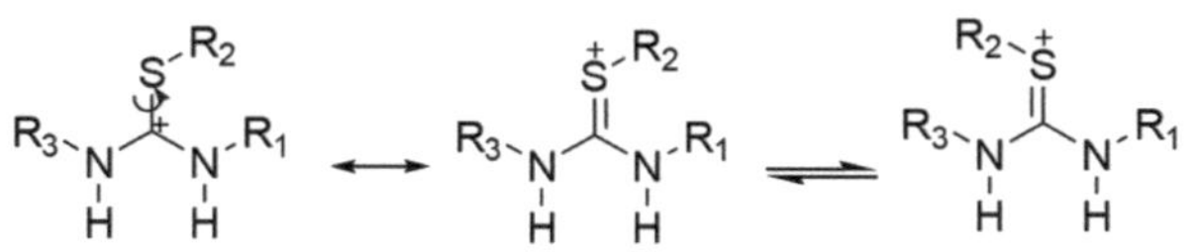

(3)

Among the above isomerisms, the *cis–trans* isomerism at the CN double bond in amides is easily detected; however, the possibility of isomerism at the CS double bond in thioketonium ions has only been examined at low temperature.

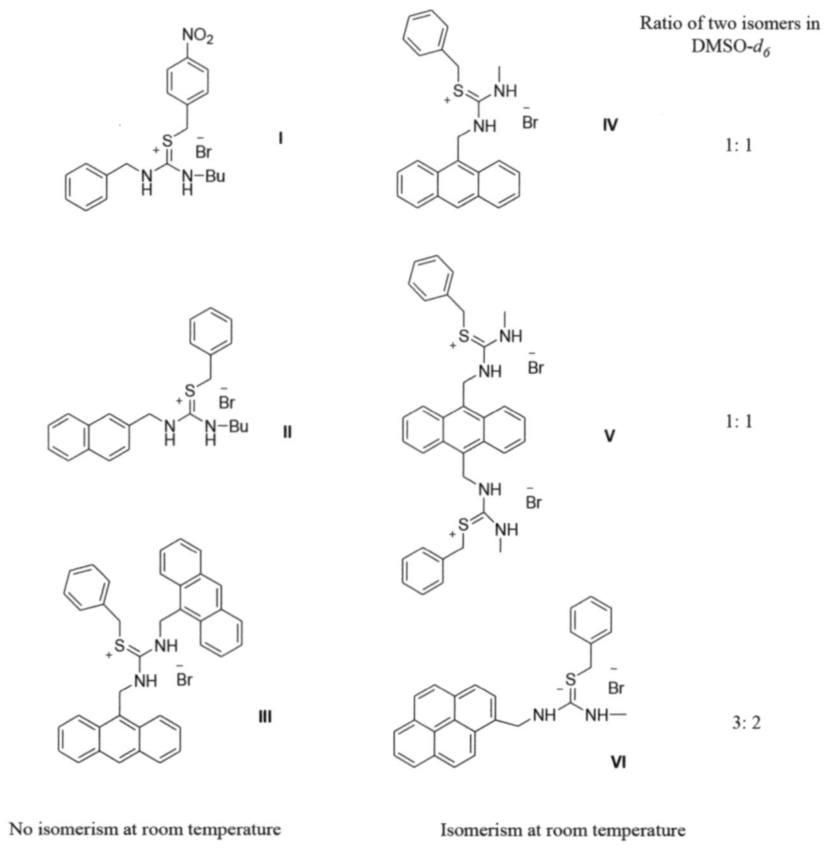

Figure 3. Isomerism in several structures of isothiouronium salts detected by NMR [14].

3. Synthetic applications of hydrogen-bonding isothiouronium organocatalysts

3.1. Nonstereoselective *S*-benzyl isothiouronium chloride organocatalyst

3.1.1. Reduction of conjugated nitroalkenes [18]

Kim et al. introduced the simple *S*-benzyl isothiouronium chloride as an efficient hydrogen-bonding organocatalyst in a series of reduction of conjugated nitroalkenes (**Scheme 2**). The mechanism of *S*-benzyl isothiouronium chloride catalyst for the activation of the reaction was assumed to be quite similar to that of the thiourea catalyst. Because of the formation of the

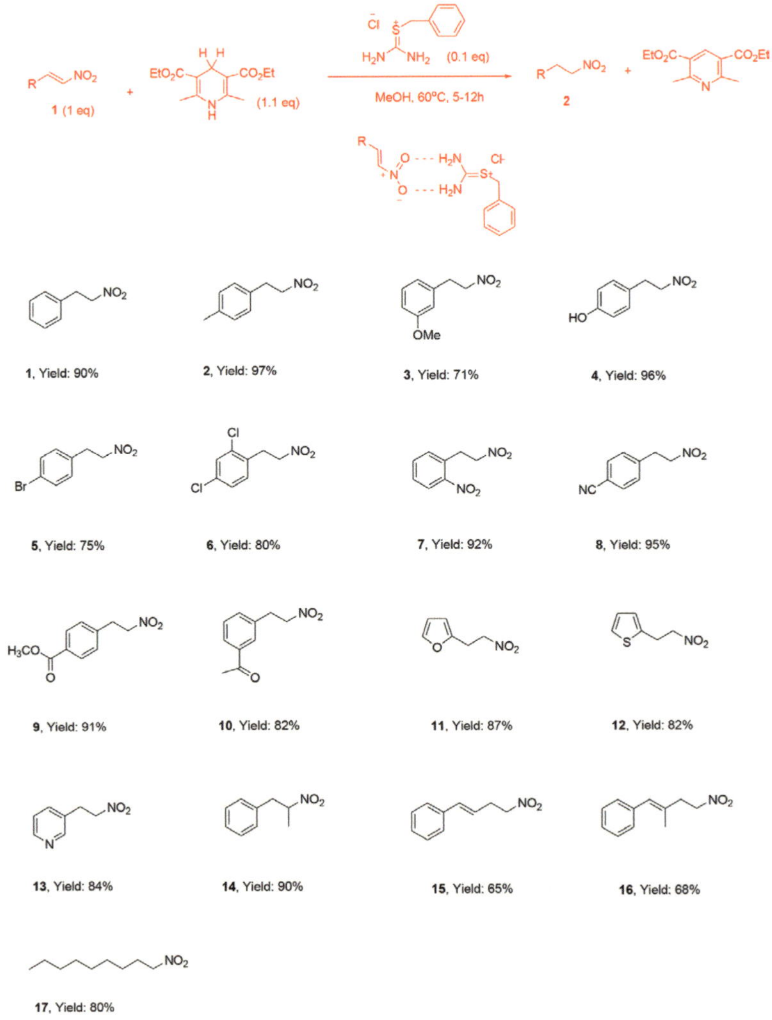

Scheme 2. Reaction scope of reduction of conjugated nitroalkenes using *S*-benzyl isothiouronium chloride catalyst.

http://dx.doi.org/10.5772/62991

strong hydrogen bonding between the nitro group and the isothiouronium moiety of the catalyst, the lowest uncccupied molecular orbital (LUMO) energy of alkenes is lower and then the reaction is accelerated. The reduction of conjugated nitroalkenes using *S*-benzyl isothiouronium chloride as a recoverable organocatalyst was successfully accomplished with high yields (65–97%) and excellent chemoselectivities. Some valuable characteristics such as the possibilities of working in the protic solvents (MeOH) and recycling (first cycle: 85%; second cycle: 88%; and third cycle: 87% yields) after converting the thiourea organocatalyst into the corresponding isothiouronium salt are the key features of this method (**Scheme 2**).

3.1.2. Reductive amination of aldehydes [19]

The authors also reported *S*-benzyl isothiouronium chloride as a novel organocatalyst with high efficiency, selectivity, and easy recovery for the direct reductive amination of aldehydes using Hantzsch ester. A mild and operationally simple fragment coupling procedure was developed, which functions with a wide range of aldehydes and amines with good-to-excellent yields (81–99%). The *S*-benzyl isothiouronium chloride catalyst can be easily recovered by simple filtration and reused with no drop in its efficiency (first cycle: 97%; second cycle: 98%; and third cycle: 95% yields). Compared with the same model based on thiourea, the isothiouronium catalyst showed higher yield and milder reaction conditions as well as easier recovery and reuse (**Schemes 3** and **4**, and **Figure 4**).

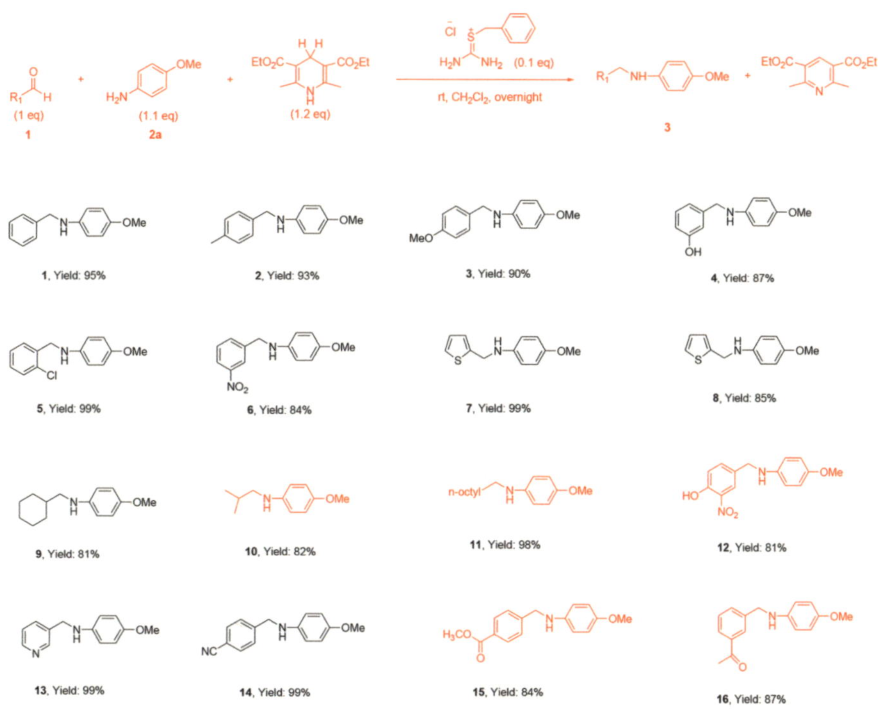

Scheme 3. Reductive amination of aldehyde scope using a variety of aldehydes.

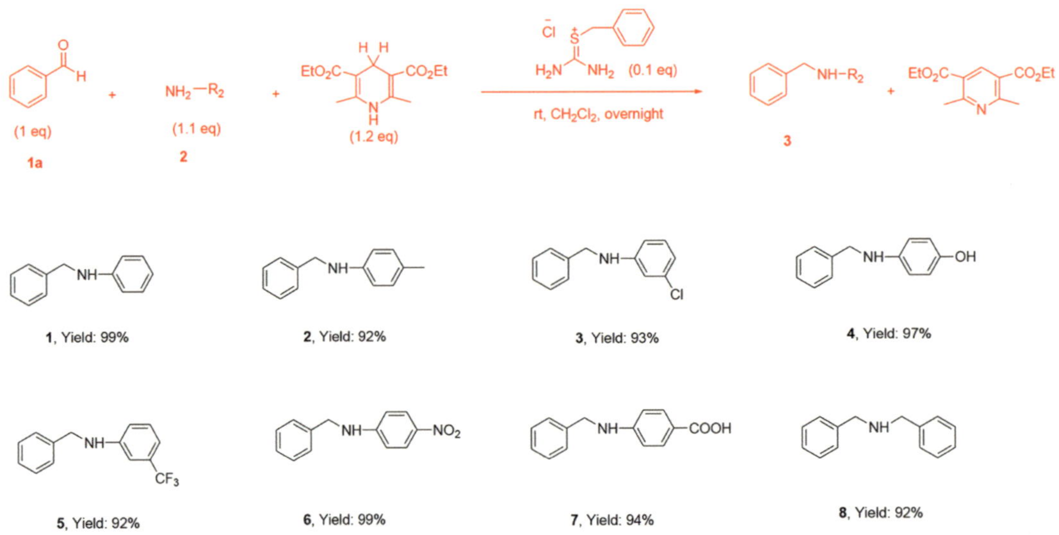

Scheme 4. Reductive amination of aldehyde scope using a variety of amines.

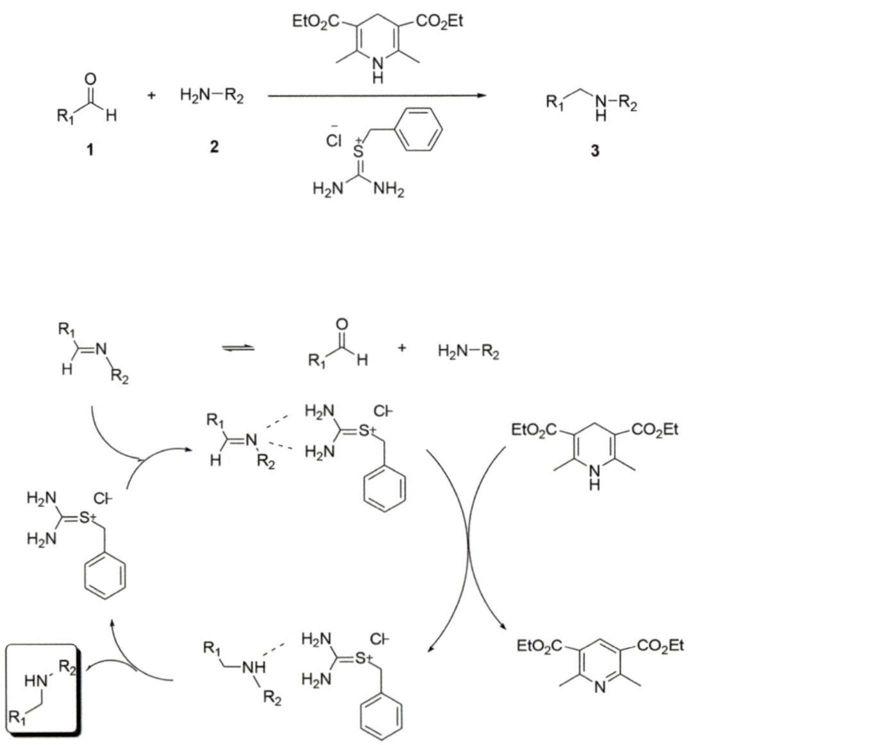

Figure 4. Proposed mechanism of the hydrogen-bond catalyzed direct reductive amination of aldehydes using *S*-benzyl isothiouronium chloride.

3.1.3. Reductive amination of ketones [20]

As part of an ongoing study of the Hantzsch ester and *S*-benzyl isothiouronium chloride system, the authors continuously reported *S*-benzyl isothiouronium chloride as a new class of noncovalent organocatalysts for the direct reductive amination of ketones (**Schemes 5** and **6**). This reaction exclusively relies on the hydrogen-bond activation by the catalyst. A wide range of ketones and amines were found to provide the expected products with moderate-to-excellent yields (45–98%). With simple modification of converting thiourea to its corresponding isothiouronium salt through alkylation, the isothiouronium catalyst acquires certain valuable characteristics such as high hydrogen-bonding propensity, the possibility of working in protic solvents (MeOH), and the ability to be recycled and reused (first cycle: 88%; second cycle: 86%; and third cycle: 85% yields).

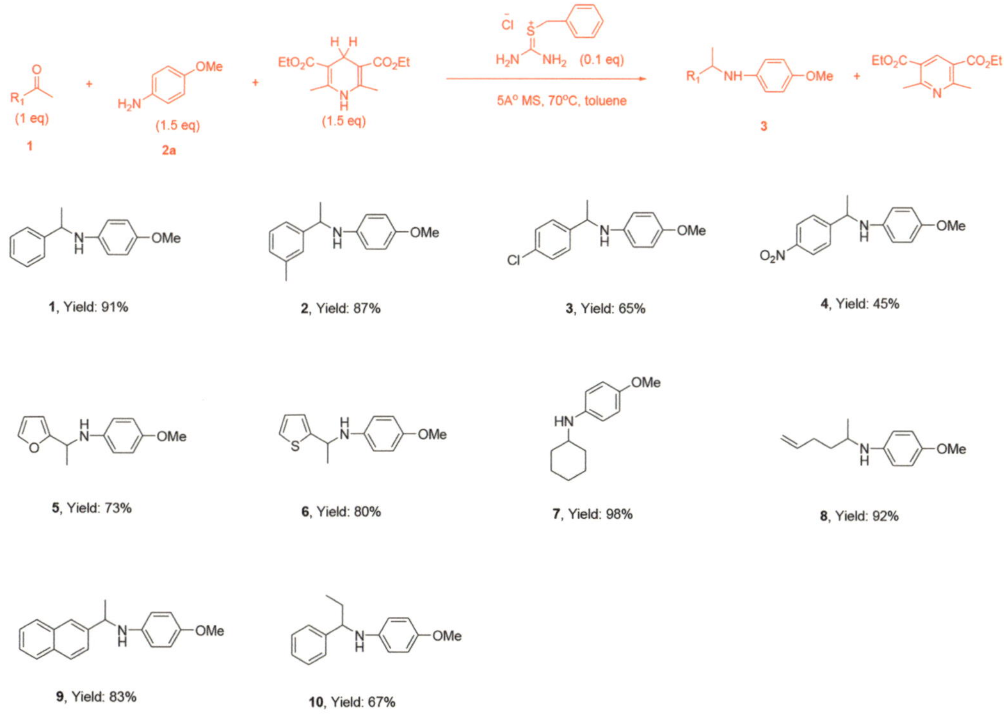

Scheme 5. Reductive amination of ketone scope using a variety of ketones.

3.2. Stereoselective isothiouronium organocatalysts

3.2.1. Enantioselective Michael addition reaction [21]

In 2009, Kilburn et al. first synthesized L-proline-based bifunctional organocatalyst containing tethered isothiouronium salt and then applied it to promote the Michael addition reaction of cyclohexanone to *trans*-β-nitrostyrene (**Scheme 7** and **Figure 5**). As compared with the same model based on thiourea, isothiouronium-substituted organocatalyst showed some enhancement of enantiocontrol (87% ee vs. 90% ee) and significant enhancement in the reaction rate (11 h vs. 5 h). Reduced catalyst loading was also tolerated (from 15 to 5%).

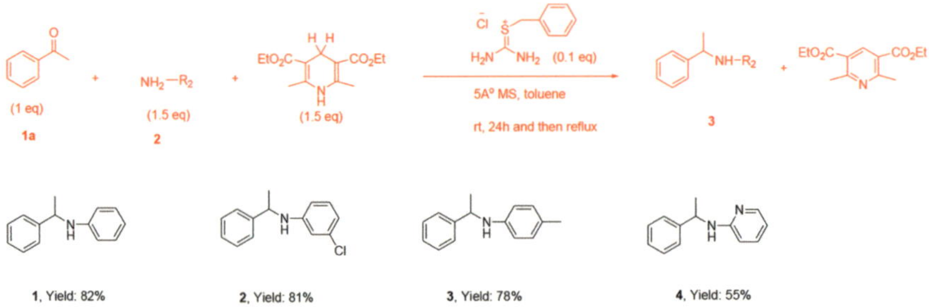

Scheme 6. Reductive amination of ketone scope using a variety of amines.

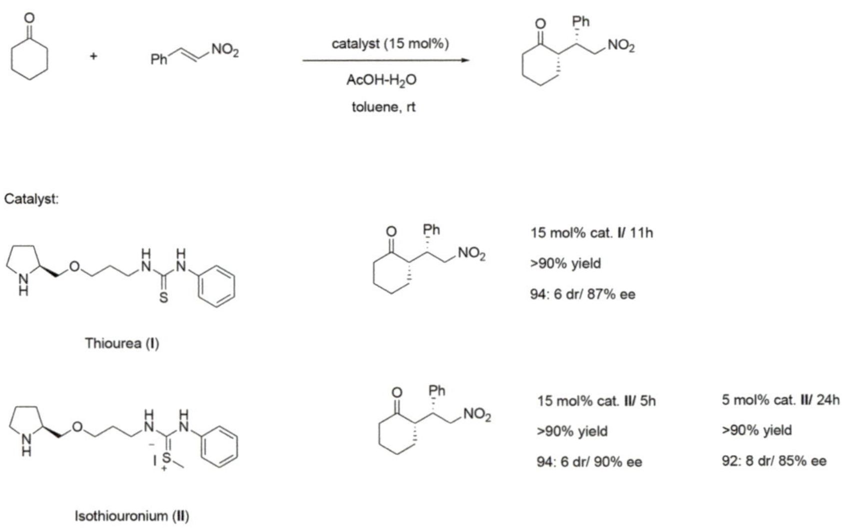

Scheme 7. Enantioselective addition of cyclohexanone to *trans*-β-nitrostyrene.

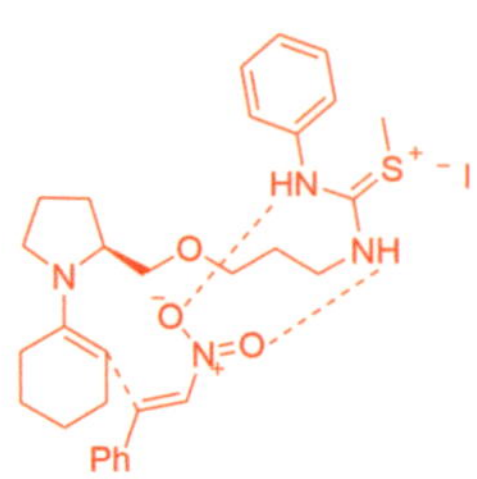

Figure 5. Proposed transition state for conjugate addition catalysis by isothiournium-functionalized organocatalyst.

http://dx.doi.org/10.5772/62991

3.2.2. Asymmetric aldol reaction [22, 23]

With a keen interest in organocatalyst systems based on isothiouronium derivatives, Kim et al. introduced isothiouronium iodide salt as an efficient cocatalyst with L-proline in the direct asymmetric aldol reactions between cyclohexanone and aromatic aldehydes [22]. This method produced good-to-excellent yields (up to 93%) with good stereoselectivities (up to 93:7 dr and 99% ee). This aldol protocol includes a solvent-free catalytic system inside a refrigerator without stirring in the hunt for an inexpensive and green process (**Table 1**, **Scheme 8**, and **Figure 6**).

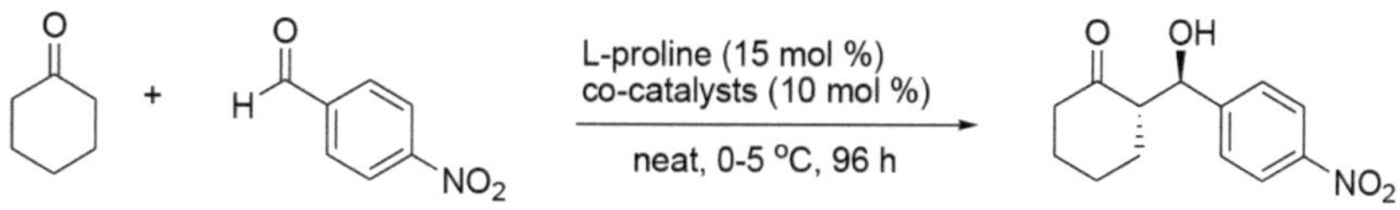

Entry	Cocatalyst	Yield (%)	*anti:syn*	ee (%)
1	None	78	81:19	56
2	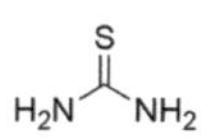	81	89:11	78
3	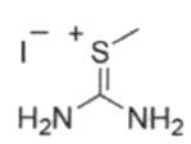	90	89:11	92
4	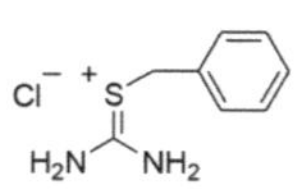	86	88:12	88
5	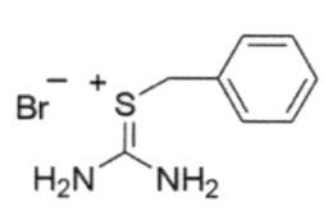	95	93:7	94
6	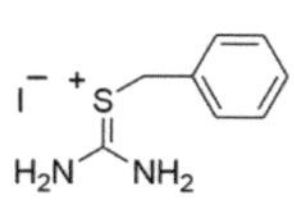	93	93:7	97
7	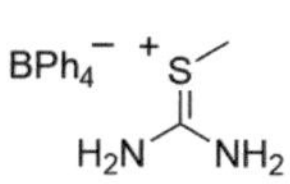	90	92:8	97

Table 1. Screening of isothiouronium cocatalysts for the L-proline-catalyzed aldol reaction.

When exploring an asymmetric aldol reaction, the authors also introduced the mimic system of isothiouronium salts which were prepared indirectly by adding acids to thiourea (**Table 2**) [23]. The isothiouronium core of salts from thiourea and acids could form a network of H-bonding interactions with the carboxylate of proline as well as with the carbonyl moieties of cyclohexanone and aromatic aldehyde, thus enhancing their electrophilicity (**Figure 7**). Asymmetric aldol reactions between cyclohexanone and aromatic aldehydes using these mimic isothiouronium systems as additives provided products with high yields (40–93%) and good stereoselectivities (up to 95:5 dr and 99% ee). This aldol protocol also enclosed a solvent-free catalytic system inside a refrigerator without stirring (**Scheme 9**).

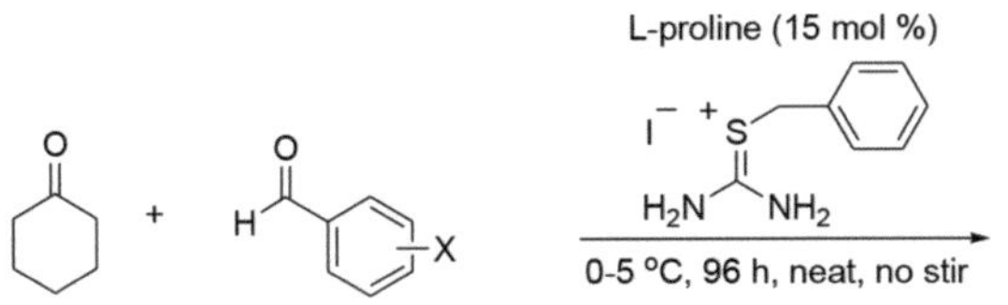

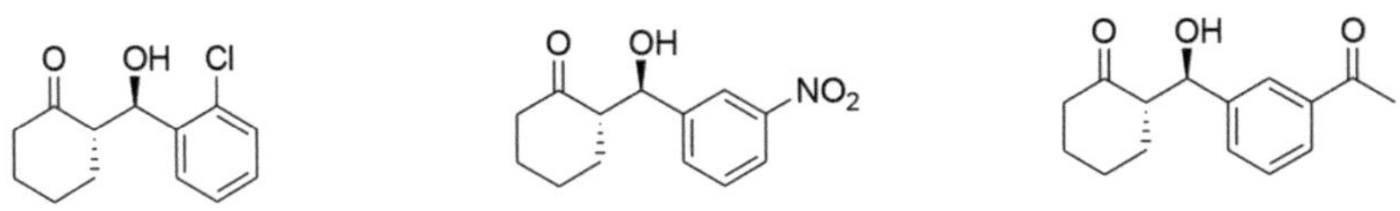

1, Yield: 90%
anti/ syn 97: 3
ee: 98%

2, Yield: 87%
anti/ syn 92: 8
ee: 91%

3, Yield: 81%
anti/ syn 93: 7
ee: 97%

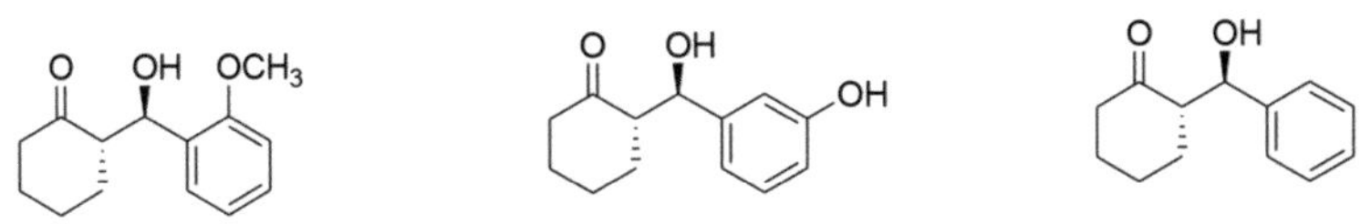

4, Yield: 37%
anti/ syn 93: 7
ee: 99%

5, Yield: 63%
anti/ syn 88: 12
ee: 94%

6, Yield: 47%
anti/ syn 89: 11
ee: 92%

Scheme 8. Asymmetric aldol reaction between various aldehydes and cyclohexanone with isothiouronium iodide.

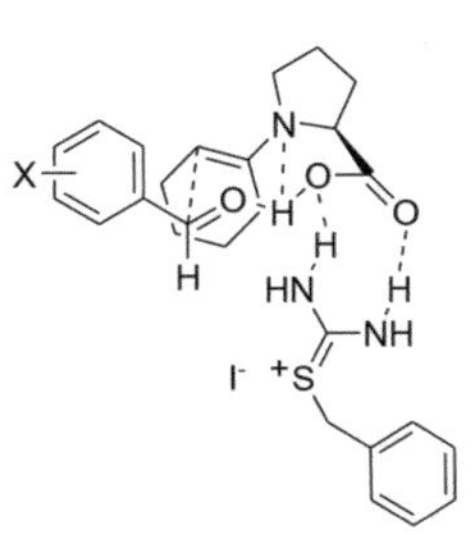

Figure 6. Proposed transition state model for asymmetric aldol reaction.

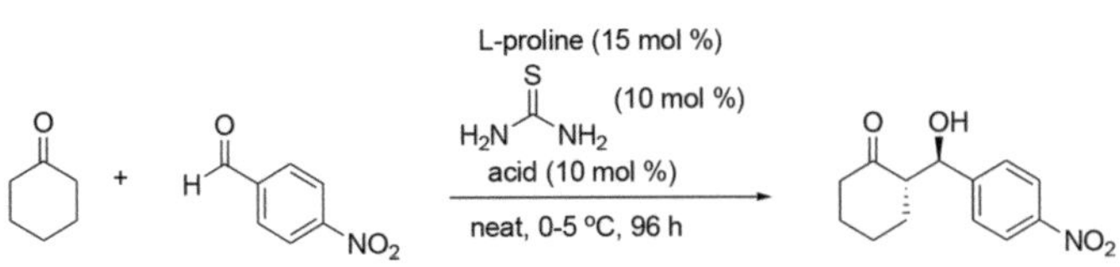

Entry	Acid	Yield (%)	*anti:syn*	ee (%)
1	None	81	89:11	78
2	None	78	81:19	56
3	$PhCO_2H$	82	93:7	92
4	CF_3CO_2H	92	95:5	98
5	HBF_4	93	95:5	99
6	HPF_6	85	96:4	99

Table 2. Screening of acids for the L-proline/thiourea-catalyzed aldol reaction.

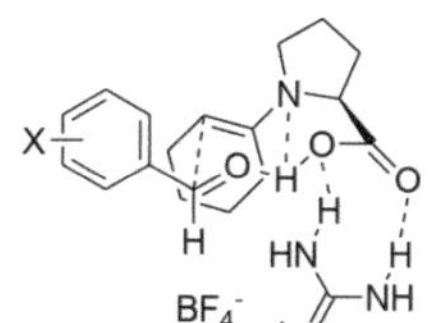

Figure 7. Proposed transition state model for asymmetric aldol reaction by adding acid to thiourea.

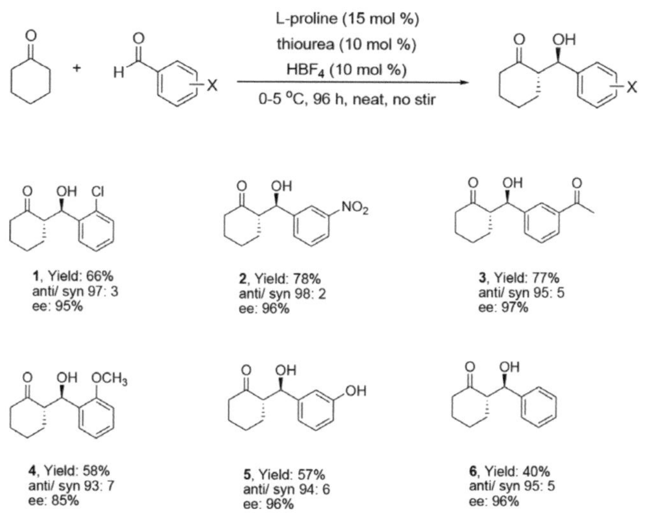

Scheme 9. Asymmetric aldol reaction with the mimic system of isothiouronium salts using thiourea/HBF_4.

4. Summary and outlook

This chapter reviews the novel research on hydrogen-bonding isothiouronium organocatalysts considering their designed concepts and synthetic applications in nonstereoselective and stereoselective reactions. From that isothiouroniums were proven to be the prospective replacements of thioureas in improving the catalytic properties because such groups would enhance the acidicity of the NH moieties for better hydrogen bonding. The milestone achievement and blooming research on thiourea catalysts have paved the avenue for further research efforts in this field. Since the chemical modification of isothiouroniums readily varies using synthetic methods to produce several types of functional molecular systems, hydrogen-bonding isothiouronium derivatives have been expected to open new approaches for the discovery of a wide range of new organocatalytic reactions.

Acknowledgements

This research was supported by the Basic Science Research Program through the National Research Foundation of Korea (NRF) funded by the Ministry of Education, Science and Technology (NRF-2013R1A1A4A01006166).

Author details

Quynh Pham Bao Nguyen and Taek Hyeon Kim*

*Address all correspondence to: thkim@jnu.ac.kr

School of Chemical Engineering, College of Engineering, Chonnam National University, Gwangju, Republic of Korea

References

[1] Walvoord RR, Huynh PNH, Kozlowski MC. Quantification of electrophilic activation by hydrogen-bonding organocatalysts. J. Am. Chem. Soc. 2014; 136: 16055-16065.

[2] MacMillan DWC. The advent and development of organocatalyst. Nature. 2008; 455: 304–308.

[3] Zhang Z, Schreiner PR. (Thio)urea organocatalysis—what can be learnt from anion recognition? Chem. Soc. Rev. 2009; 38: 1187–1198.

[4] Zabka M, Sebesta R. Experimental and theoretical studies in hydrogen-bonding organocatalysis. Molecules. 2015; 20: 15500–15524.

[5] Pihko PM (Ed.). Hydrogen bonding in organic synthesis–(thio)urea organocatalysts. Wiley-VCH: Weinheim/Germany; 2009. p. 395.

[6] Fang X, Wang CJ. Recent advances in asymmetric organocatalysis mediated by bifunctional amine-thioureas bearing multiple hydrogen-bonding donors. Chem. Commun. 2015; 51: 1185–1197.

[7] Evans NH, Beer PD. Advances in anion supramolecular chemistry: from recognition to chemical applications. Angew. Chem. Int. Ed. 2014; 53: 11716–11754.

[8] Yeo WS, Hong JI. Oxoanion recognition by a thiouronium receptor. Tetrahedron Lett. 1998; 39: 8137–8140.

[9] Kubo Y, Tsukahara M, Ishihara S, Tokita S. A simple anion chemosensor based on a naphthalene–thiouronium dyad. Chem. Commun. 2000; 653–654.

[10] Nishizawa S, Cui YY, Minagawa M, Morita K, Kato Y, Taniguchi S, Kato R, Teramae N. Conversion of thioureas to fluorescent isothiouronium-based photoinduced electron transfer sensors for oxoanion sensing. J. Chem. Soc. Perkin Trans. 2. 2002; 866–870.

[11] Kubo Y, Ishihara S, Tsukahara M, Tokita S. Isothiouronium-derived simple fluorescent chemosensors of anions. J. Chem. Soc. Perkin Trans. 2. 2002; 1455–1460.

[12] Kobo Y, Kato M, Misawa Y, Tokita S. A fluorescene-active 1,3-bis(isothiouronium)-derived naphthalene exhibiting versatile binding modes toward oxoanions in aqueous MeCN solution: new technology for sensing oxoanions. Tetrahedron Lett. 2004; 45: 3769–3773.

[13] Seong HR, Kim DS, Kim SG, Choi HJ, Ahn KH. Benzene-based tripodal isothiouronium compounds as sulphate ion receptors. Tetrahedron Lett. 2004; 45: 723–727.

[14] Nguyen QPB, Kim TH. Isothiouronium salts based on anthracene and pyrene as anion sensors. Bull. Korean Chem. Soc. 2010; 31: 712–715.

[15] Foreiter MB, Gunaratne HQN, Nockemann P, Seddon KR, Stevenson PJ, Wassell DF. Chiral thiouronium salts: synthesis, characterisation and application in NMR enantiodiscrimination of chiral oxoanions. New J. Chem. 2013; 37: 515–533.

[16] Steed JW, Gale PA. Supramolecular chemistry: from molecules to nanomaterials. John Wiley & Sons: Chichester, UK; 2012. p. 4014.

[17] Phipps RJ, Hamilton GL, Toste FD. The progression of chiral anions from concepts to applications in asymmetric catalysis. Nat. Chem. 2012; 4: 603–614.

[18] Nguyen QPB, Kim JN, Kim TH. S-Benzyl isothiouronium chloride as recoverable organocatalyst for the reduction of conjugated nitroalkenes with Hantzsch ester. Tetrahedron. 2012; 68: 6513–6516.

[19] Nguyen QPB, Kim TH. S-Benzyl isothiouronium chloride as recoverable organocatalyst for the direct reductive amination of aldehydes. Tetrahedron Lett. 2011; 52: 5004–5007.

[20] Nguyen QPB, Kim TH. S-Benzyl isothiouronium chloride as recoverable organocatalyst for the direct reductive amination of ketones with Hantzsch ester. Synthesis. 2012; 44: 1977–1982.

[21] Carley AP, Dixon S, Kilburn JD. Pyrrolidine-based organocatalysts for enantioselective Michael addition of cyclohexanone to trans-β-nitrostyrene. Synthesis. 2009; 15: 2509–2516.

[22] Cho E, Kim TH. Direct asymmetric aldol reaction co-catalyzed by L-proline and isothiouronium salts. Tetrahedron Lett. 2014; 55: 6470–6473.

[23] Cho E, Lee H, Kim TH. Asymmetric aldol reaction catalyzed by L-proline and achiral thiourea fluoroboric acid salt. Bull. Korean Chem. 2015; 36: 410–412.

[24] Sa MM, Ferreira M, Bortoluzzi AJ, Fernandes L, Cunha S. Exploring the reaction of multifunctional allylic bromides with N,S-dinucleophiles: isothiouronium salts and analogs as useful motifs to assemble the 1,3-thiazine core. Arkivoc. 2010; 11: 303–321.

[25] Nguyen QPB, Kim JN, Kim TH. Investigation of isomerism in anthracene-isothiouronium salts and application of these salts for anion sensing. Bull. Korean Chem. 2009; 30: 2093–2097.

Chapter 3

Organocatalyzed Asymmetric Reaction Using α-Isothiocyanato Compounds

Wei-Cheng Yuan, Wen-Yong Han,
Yong-Zheng Chen, Bao-Dong Cui and Xiao-Ying Xu

Additional information is available at the end of the chapter

http://dx.doi.org/10.5772/62765

Abstract

Organocatalyzed asymmetric reaction using α-isothiocyanato compounds has received much attention in the past 5 years, and significant progress has been made for three types of isothiocyanato compounds, including α-isothiocyanato amides, esters, and phosphonates. This chapter covers the recent advances of α-isothiocyanato compounds in the organocatalytic asymmetric reaction.

Keywords: Organocatalysis, Asymmetric synthesis, Enantioselectivity, α-Isothiocyanato compounds, Cascade reaction

1. Introduction

The development of efficient approaches to construct stereochemically complex compounds by catalytic asymmetric cascade reaction has received significant attention in the past 15 years [1–3]. Isothiocyanate, a new and versatile reagent for various catalyzed asymmetric cascade reactions, was firstly prepared from thiocyanate by Hofmann in 1880 [4]. However, isothiocyanato compounds, such as 3-(2-isothiocyanatoacetyl)-oxazolidin-2-one (**Figure 1**, **1a**), were not applied in catalyzed asymmetric cascade reaction until 2005 by Willis. In this process, chiral bis(oxazoline)-magnesium(II) complex was used as catalyst, delivering protected aryl β-hydroxyl-α-amino acids in good results [5]. Afterwards, different kinds of isothiocyanato compounds, including α-isothiocyanato amides, esters, and phosphonates (**Figure 1**), were developed by chemists and applied to the catalytic asymmetric syntheses [6]. In addition, the general methods for the synthesis of α-isothiocyanato compounds were shown in **Scheme 1** [5, 7–9].

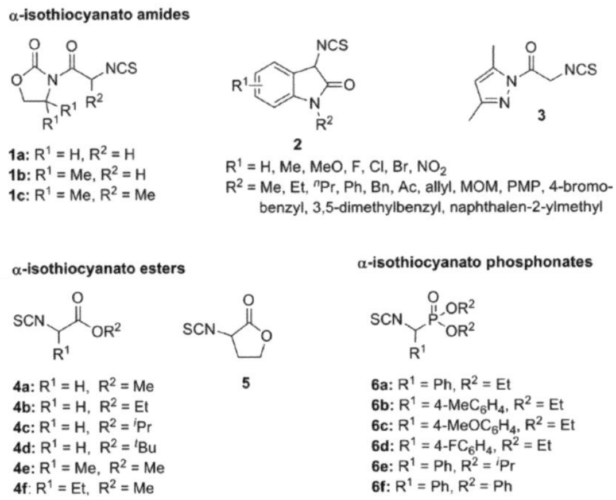

Figure 1. α-Isothiocyanato amides, esters, and phosphonates developed for the catalytic asymmetric reactions.

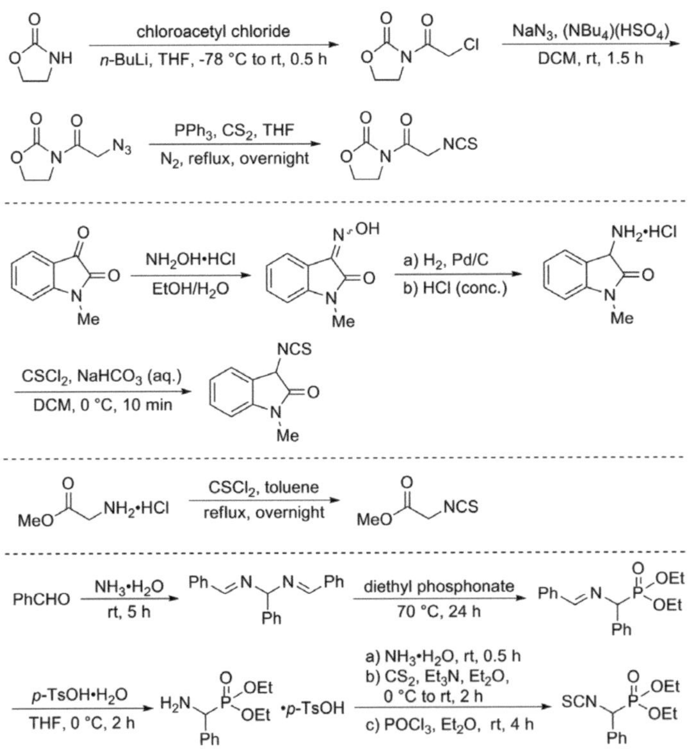

Scheme 1. Representative synthetic methods for the α-isothiocyanato compounds.

http://dx.doi.org/10.5772/62765

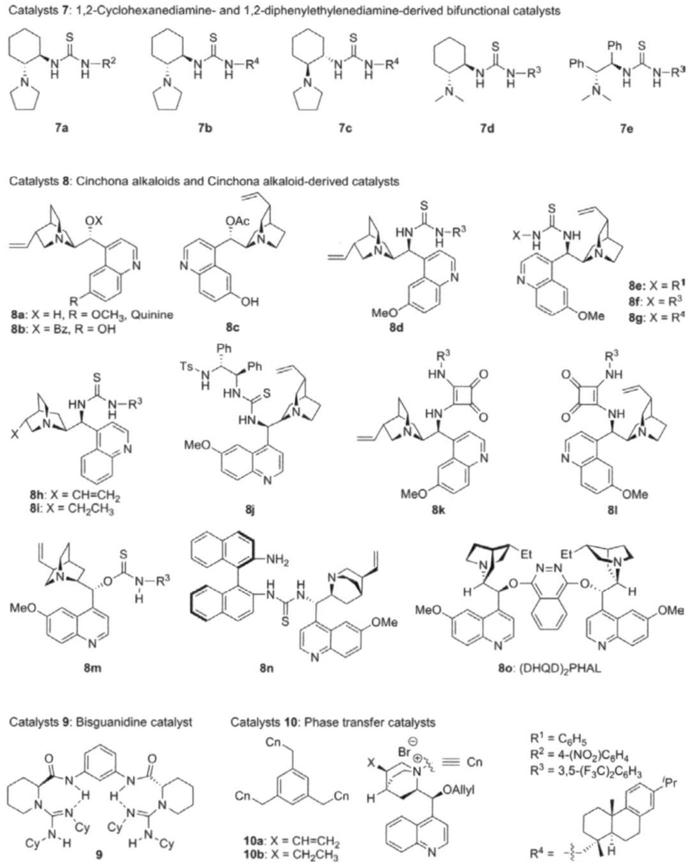

Figure 2. Organocatalysts involved in this chapter.

This chapter focuses on the synthesis of stereochemically enriched compounds based on these three kinds of isothiocyanato compounds via organocatalytic asymmetric cascade reactions, including aldol/cyclization, Mannich/cyclization, Michael/cyclization, [3+2] cyclization with allenic esters, [3+2] cyclization with 2-butynedioic acid diesters, [3+2] cyclization with azodicarboxylates, as well as self-cyclization/addition with aziridines and is classified on the basis of the types of acceptors. Some common organocatalysts involved in this chapter are listed in **Figure 2**, including 1,2-cyclohexanediamine- and 1,2-diphenylethylenediamine-derived bifunctional catalysts **7**, Cinchona alkaloids and Cinchona alkaloid-derived catalysts **8**, bisguanidine catalyst **9**, phase transfer catalysts **10**.

2. Cascade aldol/cyclization

In 2008, Seidel and coworkers successfully established the first cascade aldol/cyclization reaction of α-isothiocyanato imides (**1b**) and various aldehydes (**11**) with a chiral bifunctional thiourea–tertiary amine (**7a**) as catalyst (**Scheme 2**) [10]. With this method, a wide range of protected *syn* β-hydroxyl-α-amino acids (**12**) could be readily generated in very good results (up to 99% yield, 98:2 dr, and 96% ee) with low catalyst loading (5 mol%) and under mild reaction conditions. More importantly, in some instances, due to the product precipitation, the corresponding products even could be directly isolated with high diastereoselectivities and enantioselectivities by simple filtration.

Afterwards, another type of protected β-hydroxyl-α-amino acids was obtained in good results using the same strategy by Seidel and coworkers in 2010 (**Scheme 2**) [11]. Almost at the same time, Wang's group also reported the organocatalytic asymmetric aldol additions of α-isothiocyanato imide (**1b**) to α-ketoesters (**13b**) (**Scheme 2**) [12]. They developed the rosin-derived amine-thiourea catalyst (**7b**), and the corresponding products were obtained in good to high yields (78–99% yields) with high levels of diastereoselectivities and enantioselectivities (70:30–97:3 dr, and 81–99% ee). In comparison, Wang achieved excellent results (up to 99% ee and 97:3 dr) at very low catalyst loadings (1 mol%), but the stereocenters of bifunctional catalyst (**7a**) are more atom-economic for asymmetric synthesis.

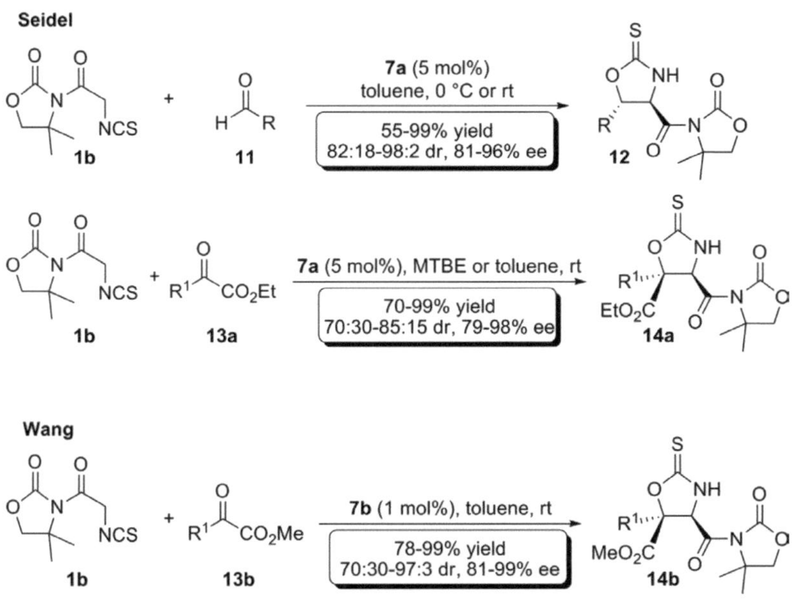

Scheme 2. Organocatalyzed aldol additions of α-isothiocyanato imides to aldehydes and α-keto esters.

Using the rosin-derived amine–thiourea **7c** as a bifunctional catalyst, a highly efficient and convenient cascade aldol/cyclization of α-isothiocyanato amides (**1b** and **1c**) or α-isothiocyanato ester (**5**) with isatins (**15**) was disclosed by Wang's group (**Scheme 3**) [13]. In this process, optically active spirocyclic thiocarbamates (**16** or **17**) were obtained in good to excellent results. Additionally, the products were readily transformed into biologically active spirooxazolines. Importantly, using a model of acute neuroinflammation, several of the spirooxazolines were found to significantly reduce LPS-induced fever.

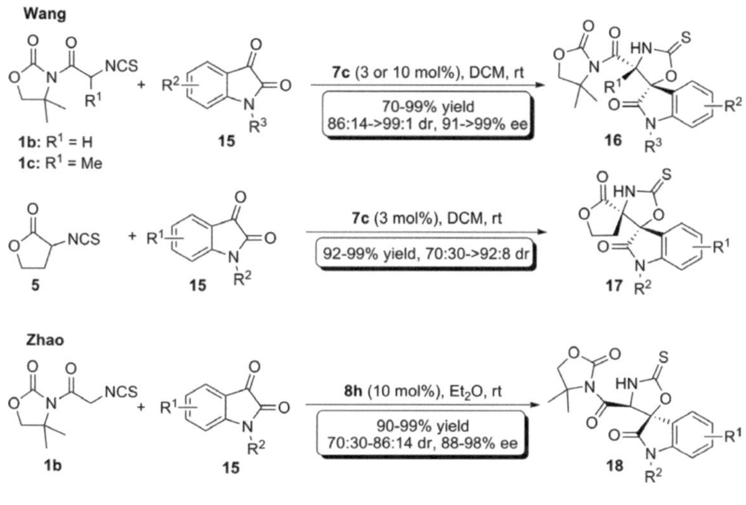

Scheme 3. Synthesis of spirocyclic thiocarbamates.

Shortly after this report, a similar protocol was reported by Zhao's group in 2011 (**Scheme 3**) [14]. Using catalyst **8h** derived from cinchonidine in diethyl ether at room temperature, a series of spirocyclic thiocarbamate compounds **18** were obtained in high yields (up to 99%) with good diastereoselectivities and enantioselectivities (up to 86:14 dr and 98% ee).

Inspired by α-isothiocyanato esters or amides are efficient nucleophilic reagents to construct chiral heterocyclics [10–14], and considering the potential applications of spirocyclic oxindole skeletons which have particularly emerged as attractive synthetic targets in the field of natural products synthesis and medicinal chemistry [15–17]. 3-Isothiocyanato oxindoles (**2**), a novel kind of α-isothiocyanato amides, were designed and synthesized by our group in 2011 [7]. Using chiral organocatalyst (**7e**), a class of 3,3′-spirooxindoles (**20**) could be formed by the reaction of 3-isothiocyanato oxindoles (**2**) with aryl and alkyl ketones (**19**). Additionally, a plausible mechanism including a sequential asymmetric aldol reaction and intramolecular cyclization of the aldol adduct with isothiocyanate unit was tentatively suggested (**Scheme 4**).

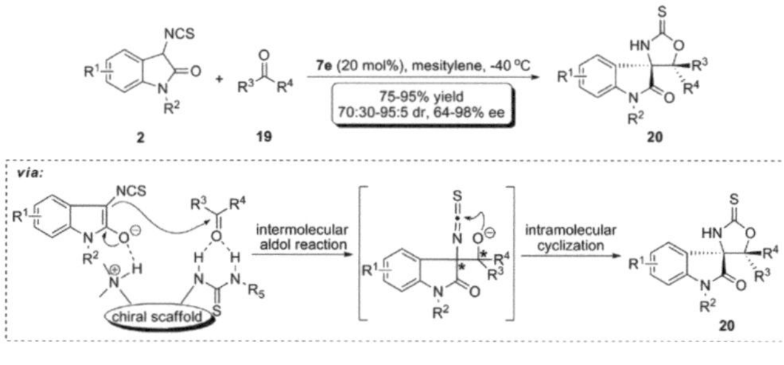

Scheme 4. Organocatalytic asymmetric aldol reactions of 3-isothiocyanato oxindoles to ketones.

Although isothiocyanates have previously been applied in the synthesis of cyclic aldol products [10–14], this represents the first example in the construction of spirocyclic oxindoles using 3-isothiocyanato oxindoles through direct aldol reaction. In the reaction of 3-isothiocyanato oxindoles (**2**) with simple ketones (**19**), a range of spirooxindole derivatives (**20**) containing two highly congested adjacent tetrasubstituted carbon stereocenters were obtained in good to excellent results (up to 95% yield, 95:5 dr, and 98% ee). Moreover, the utilities of this method were also demonstrated by the transformation of product **20a** to more structurally diverse spirooxindoles (**21–25**) (**Scheme 5**). It is worthy to note that there is no change in the stereoselectivities during the various transformations.

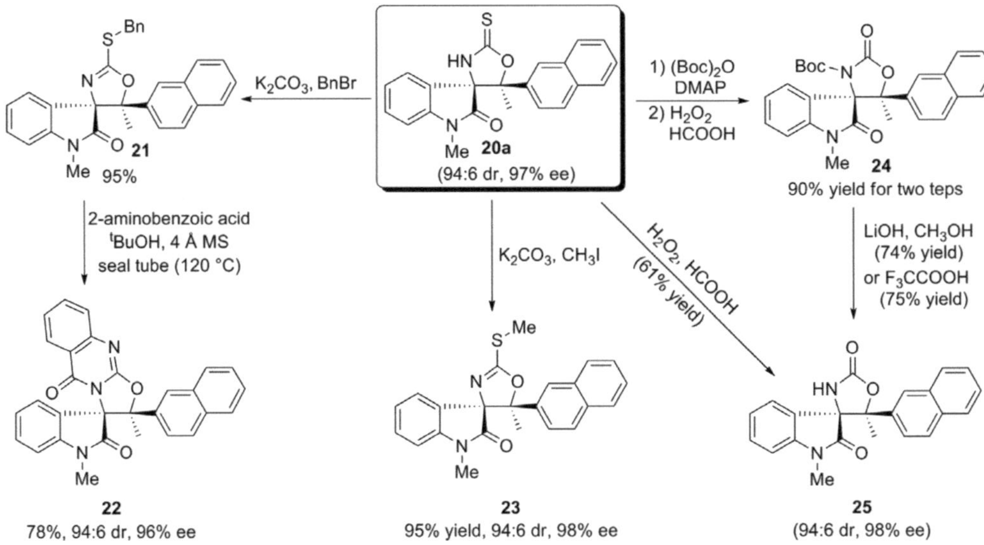

Scheme 5. Transformations of the product **20a** to other spirocyclic oxindoles.

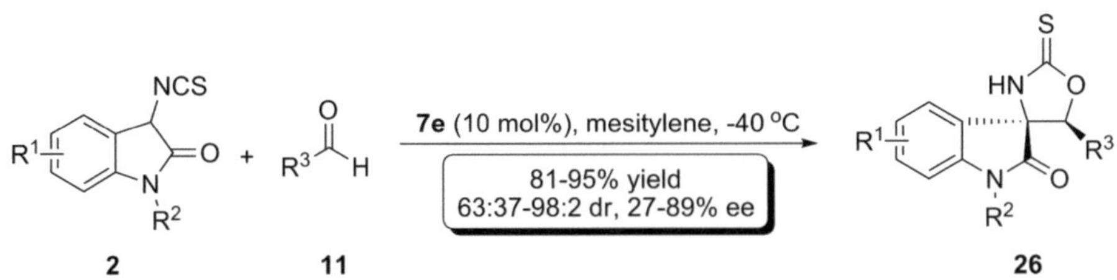

Scheme 6. Organocatalytic asymmetric aldol reactions of 3-isothiocyanato oxindoles to aldehydes.

Following this work, our group further disclosed a highly efficient approach for the synthesis of spiro[oxazolidine-2-thione-oxindoles] (**26**) with 3-isothiocyanato oxindoles (**2**) and aldehydes (**11**) via a cascade aldol/cyclization reaction process (**Scheme 6**) [18]. In this process, spiro[oxazolidine-2-thione-oxindole] compounds **26** were obtained in up to 95% yield, 98:2 dr, and 89% ee. Importantly, in most instances, high reactivity was observed and the reaction could be completed even within 1. 0 min.

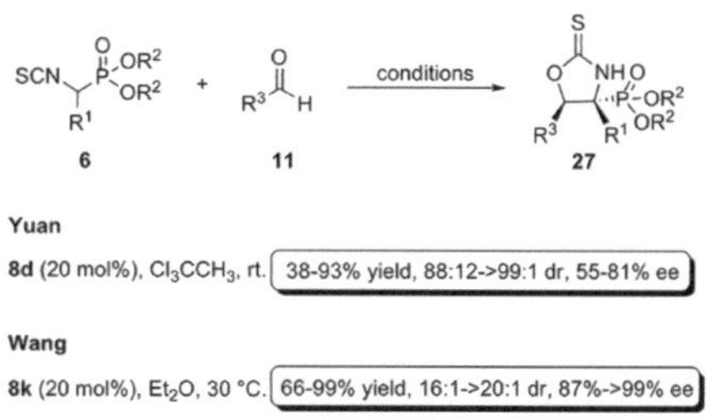

Scheme 7. Asymmetric synthesis of β-hydroxyl-α-amino phosphonic acid derivatives via organocatalytic direct aldol reaction of α-isothiocyanato phosphonates with aldehydes.

In 2013, our group firstly developed an asymmetric cascade aldol/cyclization reaction between α-isothiocyanato phosphonates (**6**) and aldehydes (**11**) for the construction of β-hydroxyl-α-amino phosphonic acid derivatives (**27**) [19]. Using bifunctional thiourea catalyst **8d** derived from quinine, a range of protected β-hydroxyl-α-amino phosphonates derivatives (**27**) containing contiguous quaternary–tertiary stereocenters was obtained in up to 93% yield, >99:1 dr, and 81% ee. The potential application of this method was demonstrated by a large- scale experiment and by the further transformation of one product.

Afterwards, a catalytic asymmetric 1,2-addition reaction of α-isothiocyanato phosphonates (**6f**) with aldehydes (**11**) was also reported from Wang's group (**Scheme 7**) [20]. Using squaramide-based hydrogen-bonding catalyst (**8k**) derived from quinine, a series of β-hydroxyl-α-amino phosphonic acids were obtained in high yields (up to 99% yield) with excellent stereoselectivities (up to >20:1 dr and >99% ee).

Recently, a novel cascade aldol/cyclization reaction of 3-isothiocyanato oxindoles with α-ketophosphonates was disclosed by Mukherjee's group (**Scheme 8**) [21]. This protocol provides an alternative approach to the β-amino-α-hydroxyphosphonate derivatives which contains a spirooxindole scaffold bearing two contiguous quaternary stereogenic centers in high yields with excellent diastereoselectivities (up to >20:1 dr) and enantioselectivities (up to >99:1 er). In addition, the products of this reaction can be modified to allow easy access to differently functionalized spirooxindoles.

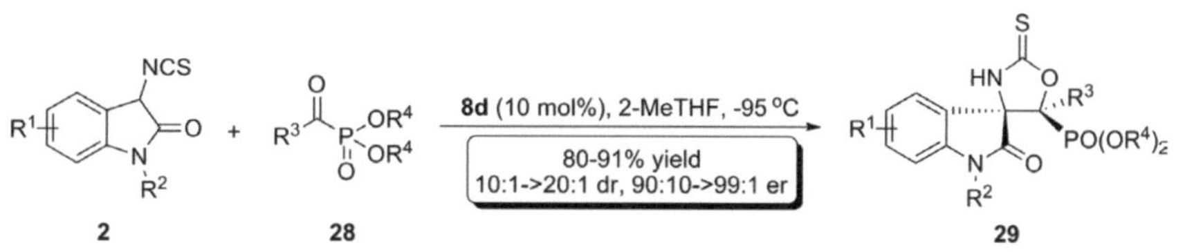

Scheme 8. Enantioselective synthesis of β-amino-α-hydroxyphosphonates between 3-isothiocyanato oxindoles and α-ketophosphonates.

3. Cascade Mannich/cyclization

Based on the remarkable success in the organocatalyzed asymmetric aldol reaction using α-isothiocyano imides (**Scheme 2**) [10], Seidel's group reported a highly stereoselective cascade Mannich/cyclization reaction between α-isothiocyanato imides (**1a**) and *N*-sulfonyl imines (**30**) in 2009 (**Scheme 9**) [22]. In this process, a diverse range of imines (Ts, Bs, and Ns) bearing substituted aromatic, heteroaromatic, and α,β-unsaturated groups could reacted with **1a** using quinidine derived **8c** as chiral catalyst, delivering *syn*-α,β-diamino acid derivatives (**31a**) in excellent results. Because of the solubility difference in the products, the reactions with Ns- and Bs-imines were much faster than Ts-imines, and the reaction of α-isothiocyanato imides with Ns- and Bs-imines carried out smoothly at a low catalyst loading (0.25–1. 0 mol%).

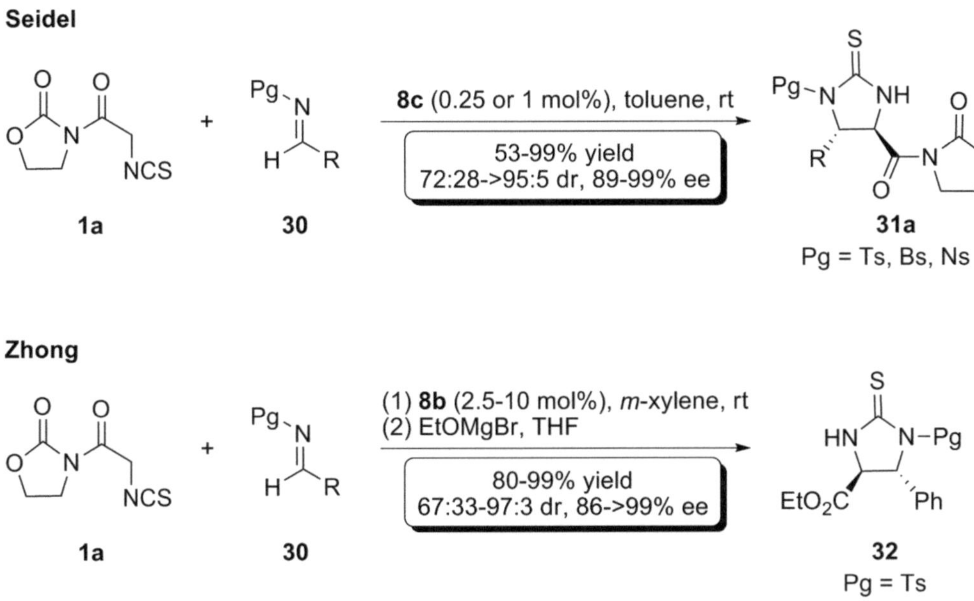

Scheme 9. Catalytic enantioselective synthesis of α,β-diamino acid derivatives.

At the same time, a similar reaction between α-isothiocyanato imide (**1a**) and Ts-imines (**30**) catalyzed by the catalyst (**8b**) derived from quinine was developed by Zhong and coworkers (**Scheme 9**) [23]. It was observed that Mannich adducts (**32**) could be achieved in good results (up to 99% yield, 97:3 dr and >99% ee).

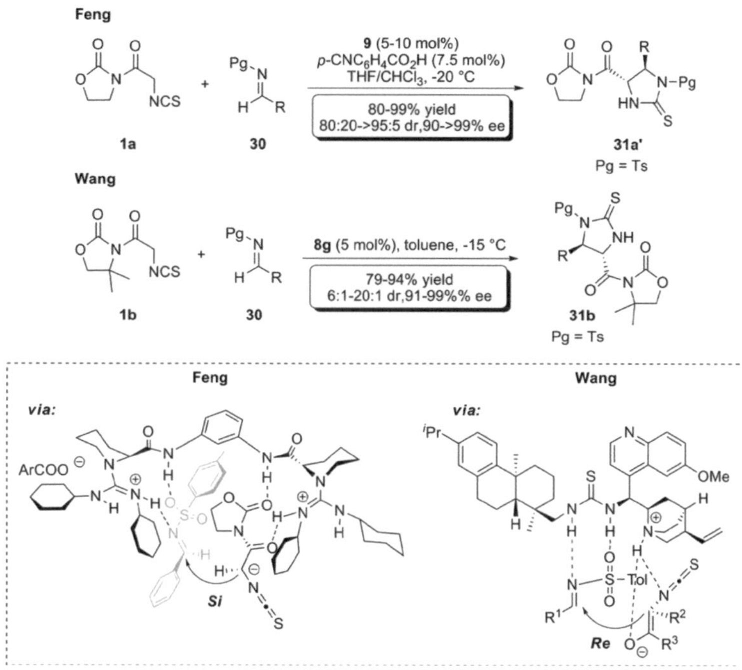

Scheme 10. Synthesis of α,β-diamino acid derivatives using bisguanidine or rosin-derived catalyst.

In 2011, Feng's group [24] and Wang's group [25] independently reported highly efficient asymmetric cascade Mannich–cyclization reaction of α-isothiocyanato imides (**1a** and **1b**) with *N*-Ts-protected imines (**30**) (**Scheme 10**). A highly efficient chiral bisguanidine catalyst (**9**) was used into the reaction of **1a** with **30** in Feng's group. In that reaction, significant progress including broad substrate scope, affording chiral α,β-diamino acid derivatives in excellent yields (up to 99%) with high diastereoselectivities (up to >95:5 dr) and excellent enantioselectivities (up to >99% ee) had been made. Notably, the additive (*p*-$CNC_6H_4CO_2H$) in the reaction had an important influence on the reactivity and asymmetric induction, which attribute to the hydrogen and the guanidine moiety both played crucial roles in the process. On the other hand, Wang's group used the bifunctional catalyst (**8g**) derived from rosin affording corresponding Mannich products **31b** in high yields (up to 94%) with good to high stereoselectivities (up to 20:1 dr, and 99% ee). And preliminary biological studies revealed that several of the methylthioimidazoline derivatives showed extremely promising antipyretic activity.

Feng and Wang also gave a proposed active model for the reaction in their work, respectively (**Scheme 10**). Feng proposed that the weakly acidic additive (*p*-$CNC_6H_4CO_2H$) could protonate one of the guanidine moieties leading to a guanidinium salt, which activates the *N*-Ts-imine via hydrogen bond to the nitrogen of the *N*-Ts-imine. Meanwhile, the active hydrogen atom of α-isothiocyanato imide was deprotonated by the other guanidine group, and then, an intermolecular hydrogen bond stabilizes the α-isothiocyanato imide. Simultaneously, the N–H group of the amide on the same side of guanidine moiety might play as a Brønsted acid to locate α-isothiocyanato amide (**1a**). In this active model, the activated α-isothiocyanato imide

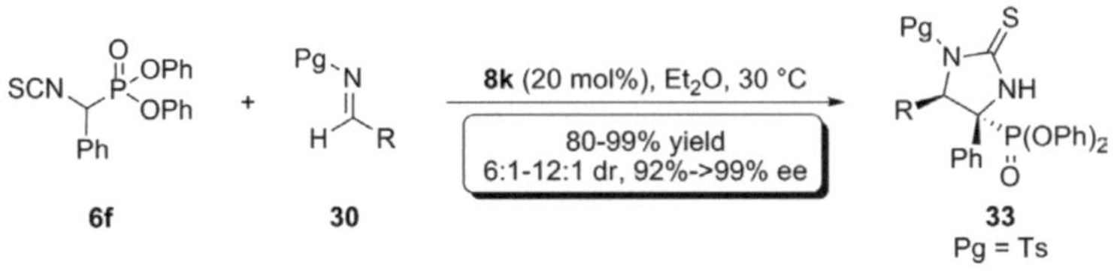

Scheme 11. Synthesis of α,β-diamino phosphonic acid derivatives.

(**1a**) preferentially attacked the activated *N*-Ts-imine from the *Si*-face, affording the desired major product with (4*S*,5*R*) configuration. On the other hand, Wang proposed that the rosin-derived tertiary amine–thiourea (**8g**) would act in a bifunctional fashion. The α-carbon atom of α-isothiocyanato amide (**1b**) was enolized by the tertiary amine moiety of the catalyst (**8g**). Concurrently, *N*-Ts-imine (**30**) was located and activated by two hydrogen atoms of the thiourea moiety through hydrogen-bond interaction. Afterwards, the nucleophile would preferentially attack at the *N*-Ts-imine (**30**) from the *Re*-face, since the *Si*-face of the *N*-Ts-imine (**30**) was restricted by the quinine moiety of the catalyst.

After the success in the asymmetric cascade aldol/cyclization reaction of diphenyl α-isothiocyanato phosphonate (**6f**) with aldehydes (**Scheme 7**) [20], the asymmetric cascade Mannich/cyclization reaction between α-isothiocyanato phosphonate (**6f**) and *N*-Ts-imine (**30**) was explored by Wang and coworkers, giving the α,β-diamino phosphonic acid derivatives (**33**) in high yields (up to 99%) with moderate diastereoselectivities and excellent enantioselectivities (up to >99% ee) (**Scheme 11**) [20]. In general, the diastereoselectivity of the cascade Mannich/cyclization reaction was lower than that of the cascade aldol/cyclization process, but the enantioselectivities of the Mannich products were slightly higher.

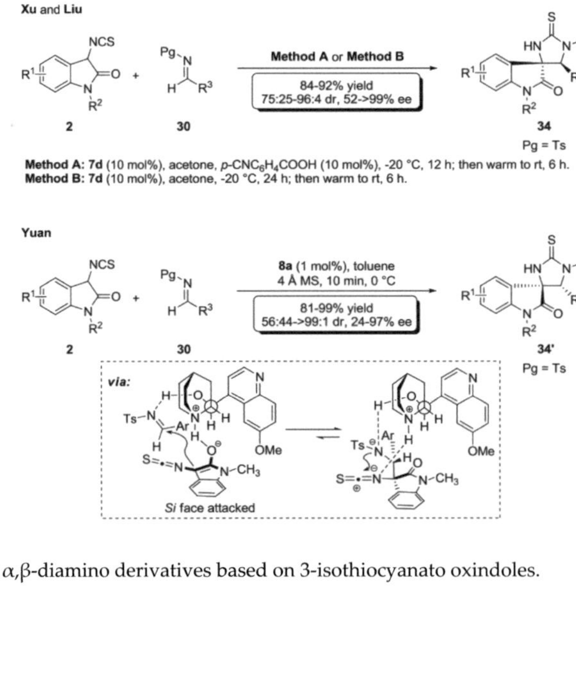

Scheme 12. Synthesis of α,β-diamino derivatives based on 3-isothiocyanato oxindoles.

Organocatalytic asymmetric cascade Mannich/cyclization reaction between 3-isothiocyanato oxindoles (**2**) and *N*-Ts imines (**30**) was reported by Xu and co-workers in 2014 (**Scheme 12**) [26]. With the developed method, the spiro[imidazolidine-4,3′-oxindole] products (**34**) were smoothly prepared in up to 92% yield, 96:4 dr, and >99% ee.

Meanwhile, a similar strategy to access spirooxindole derivatives using quinine **8a** as catalyst for the reaction of 3-isothiocyanato oxindoles (**2**) and *N*-Ts imines (**30**) was discovered in our group (**Scheme 12**) [27]. With the developed protocol, a range of structurally complex spirocyclic oxindoles derivatives (**34′**) was obtained in good results (up to 99% yield, >99:1 dr, and 97% ee). Importantly, *N*-PMP aldimine, *N*-diphenylphosphinoyl aldimine, and *N*-Boc isatinimine were also good substrates for reacting with 3-isothiocyanato oxindoles (**2**). A possible model for explaining the stereochemistry of the domino Mannich–cyclization reaction was proposed.

4. Cascade Michael/cyclization

In 2011, Wang and coworkers explored the unprecedented reaction of organocatalytic asymmetric cascade Michael/cyclization reaction using α-isothiocyanato amides [28], which

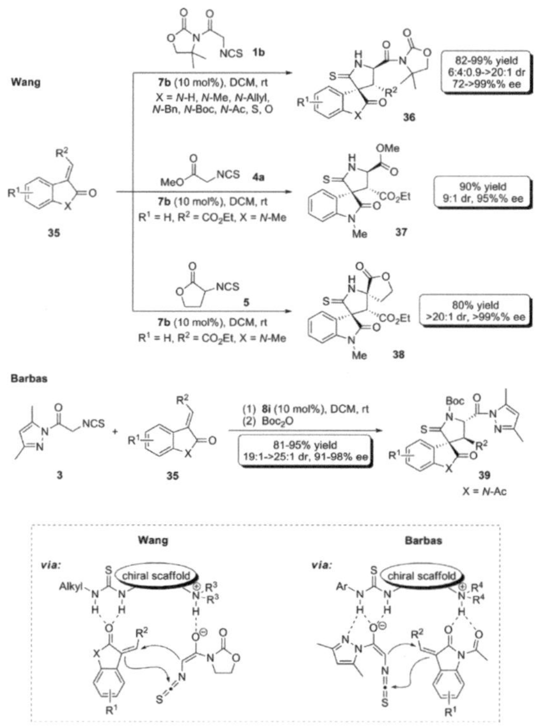

Scheme 13. Enantioselective Michael/cyclization reaction with α-isothiocyanato imide or ester.

attribute to the stereoselectivity and reactivity of the α-isothiocyanato compounds with electron-deficient olefins is challenging. As shown in **Scheme 13**, rosin-derived thiourea (**7b**) catalyzed the asymmetric cascade Michael/cyclization reaction of α-isothiocyanato imides (**1b**, **4a**, and **5**) with various methyleneindolinones (**35**), affording the densely functionalized 3,3′-pyrrolidonyl spirooxindoles (**36–38**) in excellent results (up to 99% yield, >20:1 dr, and >99% ee).

Another kind of asymmetric cascade Michael/cyclization reaction between dimethylpyrazole isothiocyanato imides (**3**) and methyleneindolinones (**35**) using bifunctional catalyst (**8i**) derived from cinchonidine was also explored by Barbas and coworkers in 2012 (**Scheme 13**) [29]. Dimethylpyrazole was selected as the directing group, because there would have interaction between the pyrazole group and the thiourea moiety of the catalyst. Additionally, the acetyl protecting group played a crucial role in the enantioselectivity. The reaction provided smoothly a range of highly functionalized spirocyclic oxindoles (**39**) in excellent results (up to 95% yield, >25:1 dr, and 98% ee). Notably, for this kind of asymmetric cascade Michael/cyclization reactions with bifunctional thiourea–tertiary amine catalyst, although two substrates were simultaneously activated by the catalyst, Wang and Barbas gave their respective transition states (**Scheme 13**). Wang proposed that the oxindole scaffold was activated by double H-bonding, and the α-carbon atom of isothiocyanate imide was activated via enolate anion by the tertiary amine moiety of the catalyst. However, Barbas thought that the oxindole scaffold was activated by the tertiary amine moiety of catalyst and that the α-carbon atom of isothiocyanato imide was deprotonated by double H-bonding.

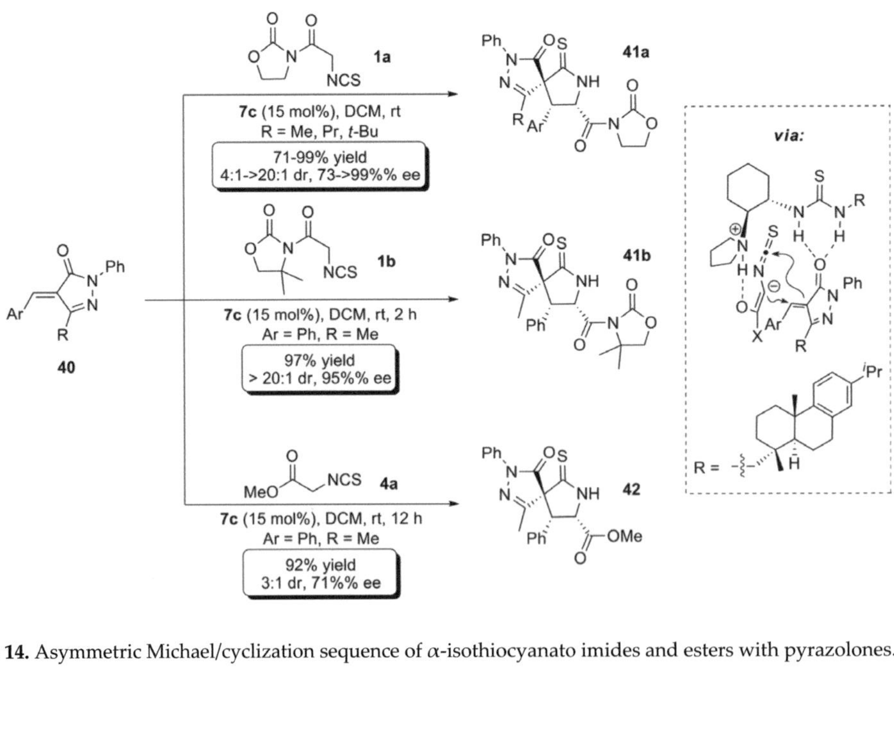

Scheme 14. Asymmetric Michael/cyclization sequence of α-isothiocyanato imides and esters with pyrazolones.

On the basis of the works from Wang's and Barbas' group (**Scheme 13**) [28, 29], organocatalytic asymmetric cascade Michael/cyclization reaction of α-isothiocyanato imides and esters (**1a**, **1b**, and **4a**) with various unsaturated pyrazolones (**40**) was also explored by Wang and coworkers in 2012 (**Scheme 14**) [30]. With the developed method, a series of spirocyclic pyrazolone compounds were observed (**41a**, **41b**, and **42**) in high diastereoselectivity and enantioselectivity (up to 20:1 dr, and >99% ee). Undoubtedly, this methodology provides a convenient and highly efficient approach for the construction of spiropyrazolone skeletons with high enantioselectivities.

3-Isothiocyanato oxindoles (**2**) are important reagents for the stereoselective construction of structurally diverse spirocyclic oxindoles via cascade aldol/cyclization process [7, 18, 21, 31] and Mannich/cyclization process [26, 27, 31]. It is worthy to note that organocatalytic asymmetric cascade Michael/cyclization reactions using 3-isothiocyanato oxindoles as donors to construct densely functionalized spirooxindoles derivatives rapidly appeared during the past four years. The following examples illustrate the progress in this area.

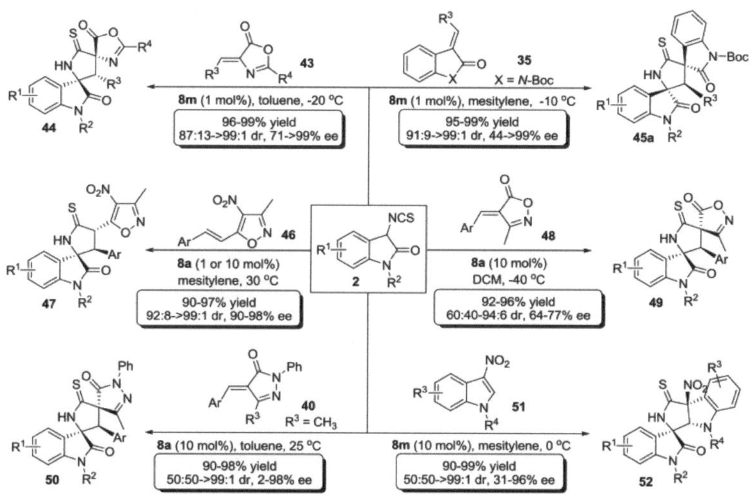

Scheme 15. Asymmetric Michael/cyclization based on 3-isothiocyanato oxindoles (**2**) by Yuan's group.

In 2013, an efficient asymmetric cascade Michael/cyclization reaction of 3-isothiocyanato oxindoles (**2**) with alkylidene azlactones (**43**) or 3-methylene-2-oxindoles (**35**) to construct more complex dispirocyclic thiopyrrolidineoxindoles was explored by our group [32], providing a range of highly functionalized dispirocyclooxindole compounds (**44** and **45a**) in excellent results (up to 99% yield, >99:1 dr, and >99% ee) only using 1 mol% of readily available aminothiocarbamate catalyst (**8m**) (**Scheme 15**).

To further investigate the versatility of 3-isothiocyanato oxindoles for constructing more complex spirocyclic oxindole scaffolds, the cascade Michael/cyclization reaction of 3-isothiocyanato oxindoles (**2**) with 3-methyl-4-nitro-5-alkenyl-isoxazoles (**46**) was subsequently explored in our laboratory (**Scheme 15**) [33]. To our delight, optically active 3,3′-pyrrolidonyl spirooxindole derivatives (**47**) could be smoothly obtained in high yields (up to 97%) with excellent stereoselectivities (up to >99:1 dr and 98% ee) with cinchona alkaloid quinine as catalyst. Nevertheless, the potential application of this method was also demonstrated by the large-scale experiment and the versatile transformation of the product into other heterocyclic compounds.

Encouraged by our these progresses using 3-isothiocyanato oxindoles (**2**) as powerful and versatile precursors in the organocatalytic asymmetric cascade Michael/cyclization reactions [32, 33], we further attempted to investigate the reactions of 3-isothiocyanato oxindoles (**2**) with unsaturated isoxazolones (**48**) and pyrazolones (**40**) (**Scheme 15**) [34]. It was observed that the reaction also worked well with quinine as catalyst under mild reaction conditions, delivering two kinds of spirooxindole derivatives (**49** and **50**) in high to excellent yields with moderate to good stereoselectivities.

Recently, an organocatalytic asymmetric Michael/cyclization cascade reaction of 3-isothiocyanato oxindoles (**2**) with 3-nitroindoles (**51**) was disclosed by our groups [35]. With the developed protocol, a wide range of polycyclic spirooxindoles (**52**), containing three contiguous chiral centers with two of them having quaternary stereocenters, could be obtained with satisfactory results (up to 99% yield, >99:1 dr, and 96% ee). This method is very promising because the reaction is scalable, and the versatile transformations of the products into other spirocyclic oxindoles are also feasible.

In 2013, catalytic asymmetric cascade Michael addition/cyclization reaction of 3-isothiocyanato oxindoles (**2**) with electron-deficient olefins (**53** or **35**) using bifunctional thiourea catalyst derived from quinidine was reported by Wang and coworkers. In their developed method, the reaction proceeded well under mild reaction conditions and afforded structurally complex 3,2′-pyrrolidinyl spirooxindole derivatives (**54**) (up to 99% yield, >20:1 dr, and 96% ee) and bispirooxindoles (**45b**) (up to 99% yield, >20:1 dr, and 98% ee), respectively, (**Scheme 16**) [36].

Having achieved remarkable success in organocatalytic asymmetric Michael/cyclization reaction with α-isothiocyanato imides and esters as versatile precursors [28, 30, 36], the asymmetric Michael/cyclization process of 3-isothiocyanato oxindoles (**2**) with unsaturated pyrazolones (**40**) using bifunctional thiourea –tertiary amines (**7c**) as catalyst was explored by Wang's group (**Scheme 16**) [37]. The reaction also provided densely functionalized spirocyclic oxindole compounds (**50**) in excellent results (up to 96% yield, >20:1 dr, and 99% ee).

In 2013, Chen and coworkers reported a formal [3+2] cycloaddition reaction between 3-isothiocyanato oxindoles (**2**) and 3-ylideneoxindoles (**35**) with quinine-derived squaramide (**8k**) as catalyst (**Scheme 16**) [38], getting multicyclic spirooxindole derivatives (**45c**) in excellent results (up to 99% yield, >95:5 dr, and >99% ee) with very low catalyst loading (1 mol%).

At the same year, a similar reaction between isothiocyanato oxindoles (**2**) and methyleneoxindoles (**35**) through a cascade Michael/cyclization process for the construction of bispirooxindole scaffolds (**45d**) was also successfully explored in Huang's group (**Scheme 16**) [39]. In this process, trifunctional organocatalyst (**8n**) based on BINAM and quinine scaffolds was efficient for the methyleneoxindole substrates bearing ketone moieties. However, bifunctional organocatalyst (**8d**) derived from quinine was the optimal catalyst for the methyleneoxindole substrates bearing different ester substituents. It should be noted that all of the reactions could complete in less than 1. 0 min, providing structurally complex bispirooxindole derivatives (**45d**) in high yields (up to 99% yield) with excellent stereoselectivities (up to >20:1 dr, and 99% ee).

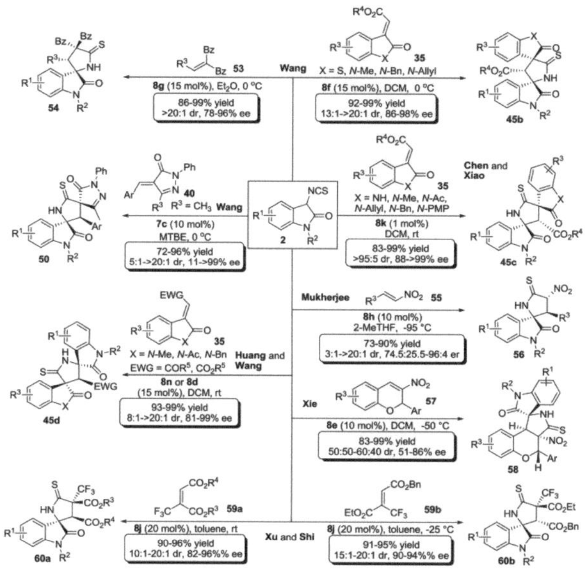

Scheme 16. Asymmetric Michael/cyclization based on 3-isothiocyanato oxindoles (**2**) by other groups.

Besides, β-substituted nitro olefins (**55**) were also good substrates for the reaction with 3-isothiocyanato oxindoles (**2**) in asymmetric Michael/cyclization process (**Scheme 16**) [40]. Structurally complex 3,2′-pyrrolidinyl-substituted spirooxindoles (**56**) were obtained in high yields (up to 90% yield) with moderate to good stereoselectivities (up to >20:1 dr, and 96:4 er).

In 2014, Xie and coworkers further developed this process by employing 3-nitro-2*H*-chromene compounds (**57**) as substrates. Their study revealed that thiourea–tertiary amine catalyst (**8e**) containing phenyl group was more efficient than the catalyst with (3,5-ditrifluoromethyl)phenyl group (**Scheme 16**) [41]. Under the optimal reaction conditions, the asymmetric cascade Michael/cyclization reaction of 3-isothiocyanato oxindoles (**2**) with 3-nitro-2*H*-chromene derivatives (**57**) proceeded smoothly, affording highly functionalized spiro[chromeno[3,4-*c*]pyrrole-1,3′-indoline] derivatives (**58**) in moderate results (up to 99% yield, 60:40 dr, and 86% ee).

The development of efficient methods to construct complex molecules containing the F atoms has been attracting considerable interest due to the potential applications of this kind of compounds in biology, medicine, and agricultural chemistry as well as in materials science. Recently, Shi's group explored a novel Cinchona alkaloid-derived multifunctional amine **8j** catalyzed asymmetric [3+2] cycloaddition of 3-isothiocyanato oxindoles (**2**) with trifluoromethylated maleates (**59a**) or fumarate (**59b**) to afford the corresponding spirooxindoles (**60**) possessing a CF_3-containing quaternary carbon stereocenter in good to excellent results (up to 96% yield, 20:1 dr, and 96% ee) (**Scheme 16**) [42]. Interestingly, two epimeric isomers were obtained with the same organocatalyst at different temperatures, which led to an enantiodivergent approach for the synthesis of spirooxindoles (**60**). From a synthetic standpoint, this study explored a new trifluoromethylation system for the synthesis of enantioenriched heterocycles with CF_3-containing quaternary carbon stereocenters.

In order to further expand the scope of α-isothiocyanato phosphonate in organocatalyzed asymmetric cascade reaction, the application of α-isothiocyanato phosphonate (**6f**) in the asymmetric Michael reaction was also investigated by Wang and coworkers [20]. Using methyleneindolinone (**35**) as Michael acceptor, the corresponding phosphonate-substituted spirooxindole (**61**) was obtained with 80% yield, >20:1 dr, and 98% ee (**Scheme 17**).

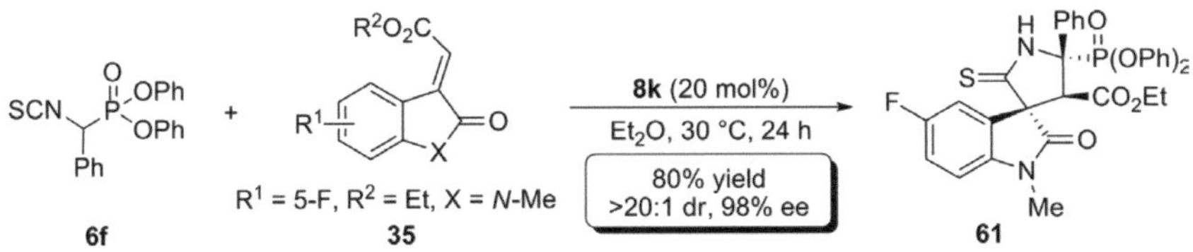

Scheme 17. Synthesis of phosphonate-substituted spirooxindole through Michael/cyclization of α-isothiocyanato phosphonate with an activated olefin.

5. Asymmetric [3+2] cyclization of 3-isothiocyanato oxindoles with allenic esters

A novel asymmetric [3+2] cycloaddition of 3-isothiocyanato oxindoles (**2**) and allenic esters (**62**) with quinine-derived organocatalyst (**8k**) was explored by Xu and coworkers in 2013 (**Scheme 18**) [43]. Interestingly, the same reaction could generate different kind of spirocyclic oxindole products (**63** or **64**) in very good results (up to 96% yield, and 97% ee) only by adjusting the ratio of the two substrates. The authors considered that organocatalyst (**8k**) having stronger hydrogen-bonding donors plays a vital structure-directing effect on the stereocontrol through the hydrogen bonds.

6. Asymmetric [3+2] cycloaddition of 3-isothiocyanato oxindoles with 2-butynedioic acid diesters

Encouraged by the achievements in the enantioselective construction of two classes of spirooxindole derivatives (**63** and **64**) (**Scheme 18**) [43], Xu and coworkers further investigated the asymmetric [3+2] cycloaddition reaction of 3-isothiocyanato oxindoles (**2**) and acetylenedicarboxylic acid diesters (**65**) for the construction of another kinds of spirooxindoles

(**Scheme 19**) [43]. Similarly, two kinds of spirocyclic oxindoles derivatives (**66** and **67**) were also respectively obtained in good to high yields with good to excellent diastereoselectivities and enantioselectivities by changing the ratio of the two substrates. Additionally, the application of this approach was also demonstrated by a large- scale experiment and by the further conversion of the cycloadduct into other more spirocyclic oxindoles.

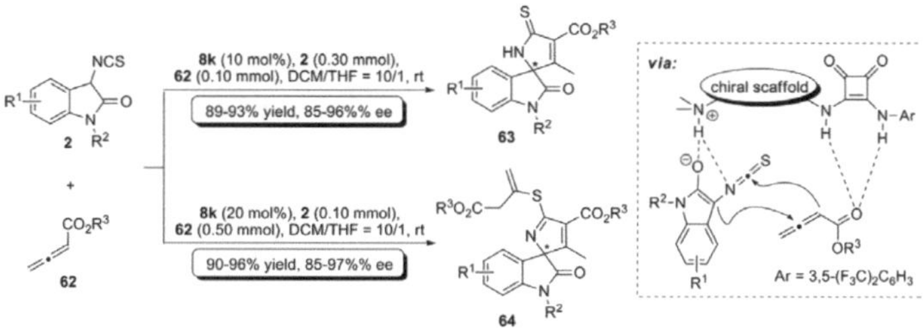

Scheme 18. Asymmetric [3+2] cycloaddition of 3-isothiocyanato oxindoles with allenic esters.

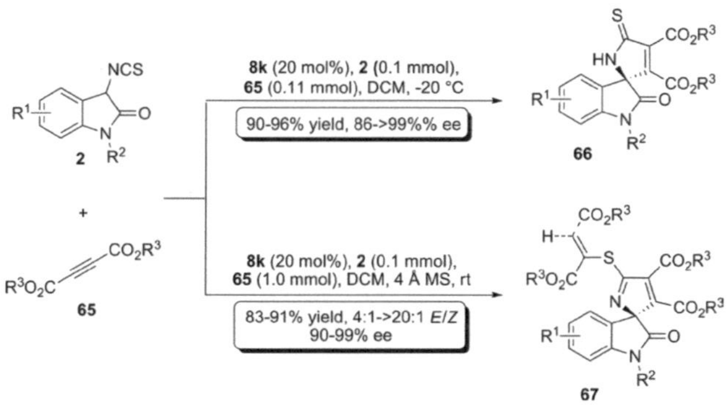

Scheme 19. Asymmetric [3+2] cyclization of 3-isothiocyanato oxindoles with 2-butynedioic acid diesters.

7. Asymmetric [3+2] cyclization of 3-isothiocyanato oxindoles with azodicarboxylates

In 2013, $(DHQD)_2PHAL$ (**8o**)-catalyzed asymmetric [3+2] cycloaddition reaction of 3-isothiocyanato oxindoles (**2**) and azodicarboxylates (**68**) was firstly explored by Xu and coworkers. In the developed method, the desired spirooxindoles (**69**) could be obtained in high yields (up to >99% yield) with high enantioselectivities (up to 98% ee) in diethyl ether with the ratio of **2** –**68** in 1.3:1 (**Scheme 20**) [44]. Similarly, another kind of spirooxindoles (**70**) could also be afforded in excellent results (up to 99% yield and 99% ee) in toluene with the ratio of **2** –**68** in 1:2. Nevertheless, various transformations of the cycloadduct into other highly functionalized spirooxindole derivatives were also realized, and some products showed promising antitumor activity via the MTT assays.

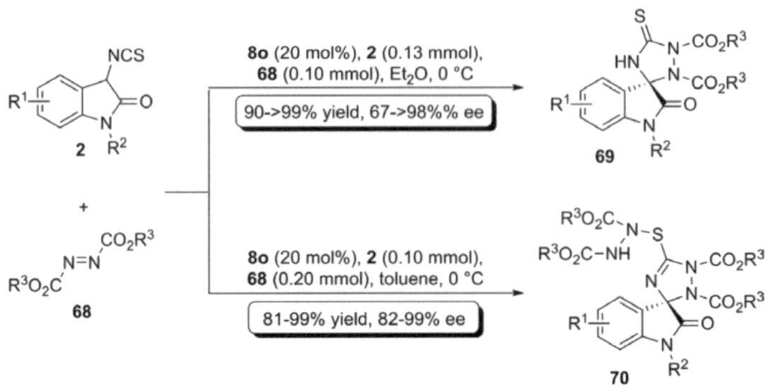

Scheme 20. Asymmetric [3+2] cyclization of 3-isothiocyanato oxindoles with azodicarboxylates.

8. Organocatalytic asymmetric ring-opening reaction of aziridines with α-isothiocyanato imides

In 2013, Wang's group developed a catalytic asymmetric ring-opening reaction of *meso*-aziridines (**71**) with α-isothiocyanato imides (**1b–c**) using trimeric quaternary ammonium salts (**10a** and **10b**) derived from cinchonine as catalysts (**Scheme 21**) [45]. This novel approach provides an efficient channel to access β-aminothiooxazole compounds (**72**) in up to 96% yield and 92% enantioselectivity.

In Wang's report [45], a plausible mechanism was proposed. As shown in **Scheme 21**, the enolized α-isothiocyanato intermediate **A** might self-cyclize by trapping the oxygen anions in the enolate. The enolized intermediate might exist in two α-isothiocyanate–thiooxazole states, such as **B** and **C** tautomerisms (**Scheme 21**). Subsequently, the sulfur-based nucleophiles in

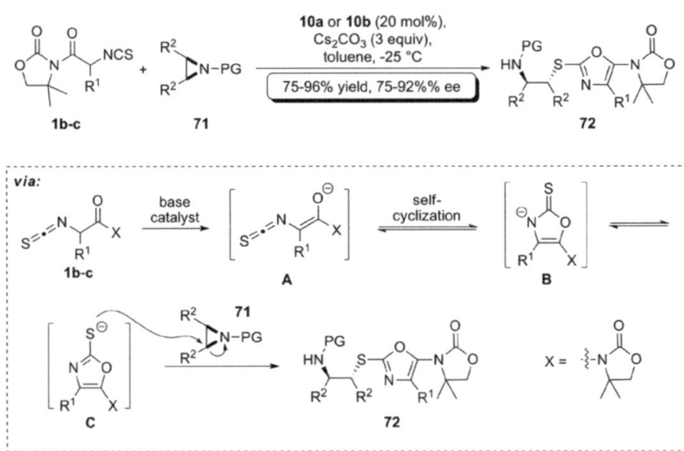

Scheme 21. Catalytic enantioselective ring-opening reaction of *meso*-aziridines with α-isothiocyanato imides.

the thiooxazole intermediate **C** attack the activated *meso*-aziridines (**71**), and consequently with the aziridine ring-opening process, leading to a series of β-aminothiooxazole compounds (**72**). Notably, this methodology extends the scope of sulfur-based nucleophiles for asymmetric ring opening of *meso*-aziridines.

9. Summary and outlook

Using α-isothiocyanato amides, esters, and phosphonates as new type of versatile reagents for the organic synthesis have been extensively studied and significant process has been made in this area over the past ten years. With these α-isothiocyanato compounds as powerful precursors, the scope of the asymmetric cascade reactions with various organocatalysts has been greatly expanded. Importantly, these developed protocols provided new access to structurally diverse and complex heterocyclic compounds which were difficult to obtain with other reactions. Despite rapid progress in organocatalytic asymmetric cascade reactions with α-isothiocyanato compounds as reagents has been made, the types of chemical reaction in this area are still limited in number. Accordingly, it might be one promising and exciting research field to develop novel asymmetric cascade reactions between the α-isothiocyanato compounds and other accepters, such as nitrosoarenes [46], *N*-sulfinylanilines [47], oxaziridines [48], nitrones [49], azomethine imines [50], α-halo *N*-acyl hydrazones [51], epoxidation [52], et al. Moreover, making further insight into the reaction mechanism concerning α-isothiocyanato reagents will facilitate the rapid development of the new methodology in this realm. In summary, the research by employing α-isothiocyanato compounds for the construction of complex heterocyclic compounds via organocatalytic asymmetric cascade process has broad development space. Further ground-breaking and exciting discoveries regarding to the application of the powerful and versatile will be reported in the near future.

Acknowledgements

The authors wish to thank the National Natural Science Foundation of China (Nos. 21372217, 21572223 and 21572224) for the financial support.

Author details

Wei-Cheng Yuan[1*], Wen-Yong Han[2], Yong-Zheng Chen[2], Bao-Dong Cui[2] and Xiao-Ying Xu[1]

*Address all correspondence to: yuanwc@cioc.ac.cn

1 Chengdu Institute of Organic Chemistry, Chinese Academy of Sciences, Chengdu, China

2 School of Pharmacy, Zunyi Medical University, Zunyi, China

References

[1] Grondal C, Jeanty M, Enders D. Organocatalytic cascade reactions as a new tool in total synthesis. Nature Chemistry. 2010; 2: 167–178. doi: 10.1038/nchem.539

[2] Volla C M R, Atodiresei I, Rueping M. Catalytic C–C bond-forming multi-component cascade or domino reactions: pushing the boundaries of complexity in asymmetric organocatalysis. Chemical Reviews. 2014; 114: 2390–2431. doi: 10.1021/cr400215u

[3] Wang Y, Lu H, Xu P-F. Asymmetric catalytic cascade reactions for constructing diverse scaffolds and complex molecules. Accounts of Chemical Research. 2015; 48: 1832–1844. doi: 10.1021/acs.accounts.5b00217

[4] Hofmann A W. Effect of high temperature on the conversion of methyl thiocyanate. European Journal of Inorganic Chemistry. 1880; 13: 1349–1352. doi: 10.1002/cber.18800130228

[5] Willis M C, Cutting G A, Piccio V J-D, Durbin M J, John M P. The direct catalytic enantioselective synthesis of protected aryl β-hydroxy-α-amino acids. Angewandte Chemie International Edition. 2005; 44: 1543–1545. doi: 10.1002/anie.200462125

[6] Han W-Y, Zhao J-Q, Zuo J, Xu X-Y, Zhang X-M, Yuan W-C. Recent advances of α-isothiocyanato compounds in the catalytic asymmetric reaction. Advanced Synthesis & Catalysis. 2015; 357: 3007–3031. doi: 10.1002/adsc.201500264

[7] Chen W-B, Wu Z-J, Hu J, Cun L-F, Zhang X-M, Yuan W-C. Organocatalytic direct asymmetric aldol reactions of 3-isothiocyanato oxindoles to ketones: stereocontrolled synthesis of spirooxindoles bearing highly congested contiguous tetrasubstituted stereocenters. Organic Letters. 2011; 13: 2472–2475. doi: 10.1021/ol200724q

[8] Zalipsky S, Albericio F, Słomczyńska U, Barany G. A convenient general method for synthesis of *Nα*- or *Nω*-dithiasuccinoyl (Dts) amino acids and dipeptides: application of polyethylene glycol as a carrier for functional purification. International Journal of Peptide and Protein Research. 1987; 30: 740–783. doi: 10.1111/j.1399-3011.1987.tb03386.x

[9] Chen M-H, Chen Z, Song B-A, Bhadury P S, Yang S, Cai X-J, Hu D-Y, Xue W, Zeng S. Synthesis and antiviral activities of chiral thiourea derivatives containing an α-aminophosphonate moiety. Journal of Agricultural and Food Chemistry. 2009; 57: 1383–1388. doi: 10.1021/jf803215t

[10] Li L, Klauber E G, Seidel D. Catalytic enantioselective aldol additions of α-isothiocyanato imides to aldehydes. Journal of the American Chemical Society. 2008; 130: 12248–12249. doi: 10.1021/ja804838y

[11] Vecchione M K, Li L, Seidel D. Catalytic enantioselective aldol additions of α-isothiocyanato imides to α-ketoesters. Chemical Communications. 2010; 46: 4604–4606. doi: 10.1039/c0cc00556h

[12] Jiang X, Zhang G, Fu D, Cao Y, Shen F, Wang R. Direct organocatalytic asymmetric aldol reaction of α-isothiocyanato imides to α-ketoesters under low ligand loading: a doubly stereocontrolled approach to cyclic thiocarbamates bearing chiral quaternary stereocenters. Organic Letters. 2010; 12: 1544–1547. doi: 10.1021/ol1002829

[13] Jiang X, Cao Y, Wang Y, Liu L, Shen F, Wang R. A unique approach to the concise synthesis of highly optically active spirooxazolines and the discovery of a more potent oxindole-type phytoalexin analogue. Journal of the American Chemical Society. 2010; 132: 15328–15333. doi: 10.1021/ja106349m

[14] Guang J, Zhao C-G. Organocatalytic enantioselective tandem aldol-cyclization reaction of α-isothiocyanato imides and activated carbonyl compounds. Tetrahedron: Asymmetry. 2011; 22: 1205–1211. doi:10.1016/j.tetasy.2011.06.028

[15] Zhou F, Liu Y-L, Zhou J. Catalytic asymmetric synthesis of oxindoles bearing a tetrasubstituted stereocenter at the C-3 position. Advanced Synthesis & Catalysis. 2010; 352: 1381–1407. doi: 10.1002/adsc.201000161

[16] Hong L, Wang R. Recent advances in asymmetric organocatalytic construction of 3,3′-spirocyclic oxindoles. Advanced Synthesis & Catalysis. 2013; 355: 1023–1052. doi: 10.1002/adsc.201200808

[17] Cheng D, Ishihara Y, Tan B, Barbas III C F. Organocatalytic asymmetric assembly reactions: synthesis of spirooxindoles via organocascade strategies. ACS Catalysis. 2014; 4: 743–762. doi: 10.1021/cs401172r

[18] Chen W-B, Han W-Y, Han Y-Y, Zhang X-M, Yuan W-C. Highly efficient synthesis of spiro[oxazolidine-2-thione-oxindoles] with 3-isothiocyanato oxindoles and aldehydes via an organocatalytic cascade aldol-cyclization reaction. Tetrahedron. 2013; 69: 5281–5286. doi: 10.1016/j.tet.2013.05.002

[19] Han W-Y, Zhao J-Q, Wu Z-J, Zhang X-M, Yuan W-C. Asymmetric synthesis of β-hydroxy-α-amino phosphonic acid derivatives via organocatalytic direct aldol reaction of α-isothiocyanato phosphonates with aldehydes. Journal of Organic Chemistry. 2013; 78: 10541–10547. doi: 10.1021/jo401779x

[20] Cao Y-M, Shen F-F, Zhang F-T, Zhang J-L, Wang R. Catalytic asymmetric 1,2-addition of α-isothiocyanato phosphonates: synthesis of chiral β-hydroxy- or β-amino-substituted α-amino phosphonic acid derivatives. Angewandte Chemie International Edition. 2014; 53: 1862–1866. doi: 10.1002/anie.201308514

[21] Kayal S, Mukherjee S. Catalytic aldol–cyclization cascade of 3-isothiocyanato oxindoles with α-ketophosphonates for the enantioselective synthesis of β-amino-α-hydroxyphosphonates. Organic Letters. 2015; 17: 5508–5511. doi: 10.1021/acs.orglett.5b02929

[22] Li L, Ganesh M, Seidel D. Catalytic enantioselective synthesis of α,β-diamino acid derivatives. Journal of the American Chemical Society. 2009; 131: 11648–11649. doi: 10.1021/ja9034494

[23] Shi Z, Yu P, Chua P J, Zhong G. Organocatalytic asymmetric synthesis of protected α,β-diamino acids. Advanced Synthesis & Catalysis. 2009; 351: 2797–2800. doi: 10.1002/adsc.200900580

[24] Chen X, Dong S, Qiao Z, Zhu Y, Xie M, Lin L, Liu X, Feng X. Guanidine organocatalyst for the asymmetric Mannich-type reaction between α-isothiocyanato imide and sulfonyl imines. Chemistry —A European Journal. 2011; 17: 2583–2586. doi: 10.1002/chem.201002571

[25] Jiang X, Wang Y, Zhang G, Fu D, Zhang F, Kai M, Wang R. Enantioselective synthesis of cyclic thioureas via Mannich reaction and concise synthesis of highly optically active methylthioimidazolines: discovery of a more potent antipyretic agent. Advanced Synthesis & Catalysis. 2011; 353: 1787–1796. doi: 10.1002/adsc.201100288

[26] Cai H, Zhou Y, Zhang D, Xu J, Liu H. A Mannich/cyclization cascade process for the asymmetric synthesis of spirocyclic thioimidazolidineoxindoles. Chemical Communications. 2014; 50: 14771–14774. doi: 10.1039/C4CC06000H

[27] Bai M, Cui B-D, Zuo J, Zhao J-Q, You Y, Chen Y-Z, Xu X-Y, Zhang X-M, Yuan W-C. Quinine-catalyzed asymmetric domino Mannich-cyclization reactions of 3-isothiocyanato oxindoles with imines for the synthesis of spirocyclic oxindoles. Tetrahedron. 2015; 71: 949–955. doi:10.1016/j.tet.2014.12.074

[28] Cao Y, Jiang X, Liu L, Shen F, Zhang F, Wang R. Enantioselective Michael/cyclization reaction sequence: scaffold-inspired synthesis of spirooxindoles with multiple stereocenters. Angewandte Chemie International Edition. 2011; 50: 9124–9127. doi: 10.1002/anie.201104216

[29] Tan B, Zeng X, Leong W W Y, Shi Z, Barbas III C F, Zhong G. Core structure-based design of organocatalytic [3+2]-cycloaddition reactions: highly efficient and stereocontrolled syntheses of 3,3′-pyrrolidonyl spirooxindoles. Chemistry : A European Journal. 2012; 18: 63–67. doi: 10.1002/chem.201103449

[30] Liu L, Zhong Y, Zhang P, Jiang X, Wang R. Core scaffold-inspired stereoselective synthesis of spiropyrazolones via an organocatalytic Michael/cyclization sequence. Journal of Organic Chemistry. 2012; 77: 10228–10234. doi: 10.1021/jo301851a

[31] Han Y-Y, Chen W-B, Han W-Y, Wu Z-J, Zhang X-M, Yuan W-C. Highly efficient and stereoselective construction of dispiro-[oxazolidine-2-thione]bisoxindoles and dispiro[imidazolidine-2-thione]bisoxindoles. Organic Letters. 2012; 14: 490–493. doi: 10.1021/ol203081x

[32] Han W-Y, Li S-W, Wu Z-J, Zhang X-M, Yuan W-C. 3-Isothiocyanato oxindoles serving as powerful and versatile precursors to structurally diverse dispirocyclic thiopyrrolidineoxindoles through a cascade Michael/cyclization process with amino-thiocarbamate catalysts. Chemistry : A European Journal. 2013; 19: 5551–5556. doi: 10.1002/chem.201300206

[33] Liu X-L, Han W-Y, Zhang X-M, Yuan W-C. Highly efficient and stereocontrolled construction of 3,3′-pyrrolidonyl spirooxindoles via organocatalytic domino Michael/cyclization reaction. Organic Letters. 2013; 15: 1246–1249. doi: 10.1021/ol400183k

[34] Cui B-D, Li S-W, Zuo J, Wu Z-J, Zhang X-M, Yuan W-C. Quinine-catalyzed asymmetric domino Michael-cyclization reaction for the synthesis of spirocyclic oxindoles bearing two spiro quaternary centers and three consecutive stereocenters. Tetrahedron. 2014; 70: 1895–1902. doi:10.1016/j.tet.2014.01.036

[35] Zhao J-Q, Zhou M-Q, Wu Z-J, Wang Z-H, Yue D-F, Xu X-Y, Zhang X-M, Yuan W-C. Asymmetric Michael/cyclization cascade reaction of 3-isothiocyanato oxindoles and 3-nitroindoles with amino-thiocarbamate catalysts: enantioselective synthesis of polycyclic spirooxindoles. Organic Letters. 2015; 17: 2238–2241. doi: 10.1021/acs.orglett.5b00850

[36] Cao Y-M, Shen F-F, Zhang F-T, Wang R. Catalytic asymmetric Michael addition/cyclization of isothiocyanato oxindoles: highly efficient and versatile approach for the synthesis of 3,2′-pyrrolidinyl mono- and bi-spirooxindole frameworks. Chemistry : A European Journal. 2013; 19: 1184–1188. doi: 10.1002/chem.201204114

[37] Chen Q, Liang J, Wang S, Wang D, Wang R. An asymmetric approach toward chiral multicyclic spirooxindoles from isothiocyanato oxindoles and unsaturated pyrazolones by a chiral tertiary amine thiourea catalyst. Chemical Communications. 2013; 49: 1657−1659. doi: 10.1039/c3cc38386e

[38] Tan F, Cheng H-G, Feng B, Zou Y-Q, Duan S-W, Chen J-R, Xiao W-J. Highly enantioselective organocatalytic Michael addition/cyclization cascade reaction of ylideneoxindoles with isothiocyanato oxindoles: A formal [3+2] cycloaddition approach to optically active bispirooxindole derivatives. European Journal of Organic Chemistry. 2013; 2013: 2071–2075. doi: 10.1002/ejoc.201300081

[39] Wu H, Zhang L-L, Tian Z-Q, Huang Y-D, Wang Y-M. Highly efficient enantioselective construction of bispirooxindoles containing three stereocenters through an organocatalytic cascade Michael–cyclization reaction. Chemistry : A European Journal. 2013; 19: 1747–1753. doi: 10.1002/chem.201203221

[40] Kayal S, Mukherjee S. Catalytic asymmetric Michael addition/cyclization cascade reaction of 3-isothiocyanatooxindoles with nitro olefins. European Journal of Organic Chemistry. 2014; 2014: 6696–6700. doi: 10.1002/ejoc.201402534

[41] Fu Z-K, Pan J-Y, Xu D-C, Xie J-W. Organocatalytic domino Michael/cyclization reaction: efficient synthesis of multi-functionalized tetracyclic spirooxindoles with multiple stereocenters. RSC Advances. 2014; 4: 51548–51557. doi: 10.1039/c4ra07860h

[42] Du D, Jiang Y, Xu Q, Tang X-Y, Shi M. Enantioselective [3+2] cyclization of 3-isothiocyanato oxindoles with trifluoromethylated 2-butenedioic acid diesters. ChemCatChem. 2015; 7: 1366–1371. doi: 10.1002/cctc.201500141

[43] Du D, Jiang Y, Xu Q, Shi M. Enantioselective construction of spirooxindole derivatives: asymmetric [3+2] cyclization of isothiocyanatooxindoles with allenic esters or 2-butynedioic acid diesters. Advanced Synthesis & Catalysis. 2013; 355: 2249–2256. doi: 10.1002/adsc.201300460

[44] Jiang Y, Pei C-K, Du D, Li X-G, He Y-N, Xu Q, Shi M. Enantioselective synthesis of spirooxindoles: asymmetric [3+2] cycloaddition of (3-isothiocyanato)oxindoles with azodicarboxylates. European Journal of Organic Chemistry. 2013; 2013: 7895–7901. doi: 10.1002/ejoc.201301418

[45] Cao Y-M, Zhang F-T, Shen F-F, Wang R. Catalytic enantioselective ring-opening reaction of *meso*-aziridines with α-isothiocyanato imides. Chemistry : A European Journal. 2013; 19: 9476–9480. doi: 10.1002/chem.201300297

[46] Pous J, Courant T, Bernadat G, Iorga B I, Blanchard F, Masson G. Regio-, diastereo-, and enantioselective nitroso-Diels–Alder reaction of 1,3-diene-1-carbamates catalyzed by chiral phosphoric acids. Journal of the American Chemical Society. 2015; 137: 11950–11953. doi: 10.1021/jacs.5b08515

[47] Jian T-Y, He L, Tang C, Ye S. N-heterocyclic carbene catalysis: enantioselective formal [2+2] cycloaddition of ketenes and *N*-sulfinylanilines. Angewandte Chemie International Edition. 2011; 50: 9104–9107. doi: 10.1002/anie.201102488

[48] Williamson K S, Michaelis D J, Yoon T P. Advances in the chemistry of oxaziridines. Chemical Reviews. 2014; 114: 8016–8036. doi: 10.1021/cr400611n

[49] Anderson L L. Diverse applications of nitrones for the synthesis of heterocyclic compounds. Asian Journal of Organic Chemistry. 2016; 5: 9–30. doi: 10.1002/ajoc.201500211

[50] Nájera C, Sansano J M, Yus M. 1,3-Dipolar cycloadditions of azomethine imines. Organic & Biomolecular Chemistry. 2015; 13: 8596–8636. doi: 10.1039/c5ob01086a

[51] Attanasi O A, Crescentini L D, Favi G, Mantellini F, Mantenuto S, Nicolini S. Interceptive [4+1] annulation of in situ generated 1,2-diaza-1,3-dienes with diazo esters: direct access to substituted mono-, bi-, and tricyclic 4,5-dihydropyrazoles. Journal of Organic Chemistry. 2014; 79: 8331–8338. doi: 10.1021/jo5016097

[52] Chen S, Mercado B Q, Bergman R G, Ellman J A. Regio- and diastereoselective synthesis of highly substituted, oxygenated piperidines from tetrahydropyridines. Journal of Organic Chemistry. 2015; 80: 6660–6668. doi: 10.1021/acs.joc.5b00816

Chapter 4

Recent Advances in Guanidine-Based Organocatalysts in Stereoselective Organic Transformation Reactions

Shrawan Kumar Mangawa and
Satish Kumar Awasthi

Additional information is available at the end of the chapter

http://dx.doi.org/10.5772/63268

Abstract

Tremendous efforts have been put toward the design and synthesis of newer enantioselective organocatalysts for the enanatioselective synthesis. Recently, guanidine-containing chiral organocatalysts have attracted considerable attention due to their ease of synthesis and high enantioselective catalytic activities. This chapter highlights the successive development of chiral guanidine organocatalysts in asymmetric organic transformation reactions in the past few decades.

Keywords: asymmetric organocatalysis, biologically active molecule, stereoselective organic transformation, guanidine group, asymmetric reactions

1. Introduction

Synthesis of enantiomerically pure molecules having multiple chiral centers is one of the ultimate goals in organic chemistry due to their importance in pharmaceutical science. It led to the development of stereospecific reactions, the most challenging fields in organic chemistry. As a result, asymmetric organocatalysts have become an interesting research field for chiral molecule synthesis. Small organic molecules have versatile functions such as efficient and selective catalytic properties that attribute toward their important roles in the construction of complex and enantiopure molecular skeletons [1].

Indeed, catalytic asymmetric inductions were successfully achieved in the second half of the twentieth century by employing transition metal catalysts [2]. Enantioselective C–C bonds and C–heteroatom bond formations have since been demonstrated by numerous research groups worldwide using the power of transition metals. Chiral ligands (organic molecule) form complexes with transition metals such as palladium, ruthenium, and rhodium that provide necessary chiral environment for an asymmetric induction. The versatility of transition metal complex was continually explored for the development of interesting methodologies in the chiral transformation reactions [3]. The quote *"Organic Synthesis—Where now?"* given by Professor Dieter Seebach, often cited from his famous review article, is so promising in the field of transition metal catalysis [4].

The experimental conditions for transition metal catalysts have been proved challenging despite the promising catalysts. One of the greatest challenges of transition metal catalysis is that such reactions required very stringent conditions such as rigorous Schlenk and degassing techniques or preparation and reaction in glove boxes. Moreover, these catalysts are often air and moisture sensitive which pose problems especially in their long-term storage and handling.

In the turn of twentieth century, the historical landmark in the field of asymmetric catalysis was witnessed with the onset of asymmetric organocatalysis. Chiral motifs bearing organic molecules derived from nature's chiral pool such as amino acids have been designed and used in catalytic amounts for enantioselective bond formation.

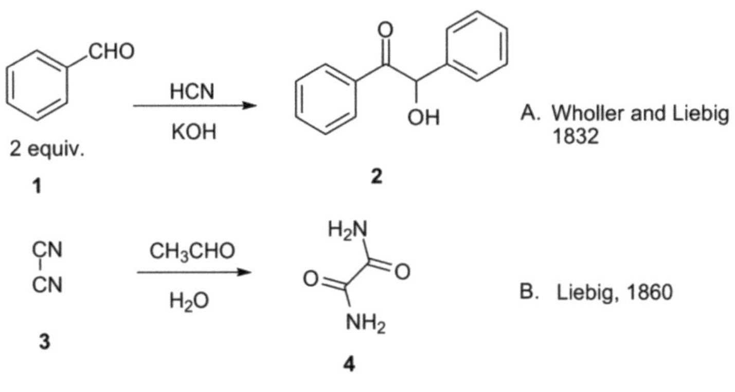

Figure 1. Organocatalyzed reactions.

Wohler and Liebig first time reported organocatalyzed Benzoin reaction [5] for the formation of the α-hydroxyl ketone in the presence of cyanide organocatalyst using two equivalents of benzaldehyde (**Figure 1A**). Liebig [6] in 1860 synthesized oxamide in the presence of acetaldehyde as organocatalyst from dicyn and water (**Figure 1B**). Another organocatalyzed reaction was the Knoevenagel condensation achieved by Emil Knoevenagel [7] in 1896 by reacting dimethyl malonate with benzaldehyde using piperidine as organocatalyst to generate the condensation product.

Subsequently, later in the turn of the twentieth century, some sporadic reports of organocatalytic reactions came up. Bredig and Fiske [8] used cinchona alkaloid as organocatalyst for asymmetric addition of HCN to benzaldehyde with low enantioselectivity (10%). Later, in 1960 Pracejus [9] reported chiral cinchona alkaloid-catalyzed methanolysis of a ketene with

http://dx.doi.org/10.5772/63268

moderate enantioselectivity (74%). Pracejus used the German terminology "Organische Katalysatoren" which nowa days is used by scientists as the term "organocatalysis." Sheehan et al. [10] in 1966 first time used *N*-heterocyclic carbene as organocatalyst. The major breakthrough in the field of organocatalyst research field was seen in 1971 and 1974, when L-proline was used in the aldol condensation reaction with excellent enantioselectivity and this reaction is better known as Hajos-Parrish-Eder-Sauer-Wiechert reaction (**Scheme 1**) [11]. Subsequently, there was no activity in the organocatalysis research field in the next two decades despite the pioneering efforts by industrial chemists in organocatalysis. At the turn of the twenty-first century, List et al. [12] reinvestigated the asymmetric organocatalyzed intermolecular aldol reaction of the Hajos-Parrish-Eder-Sauer-Wiechert reaction.

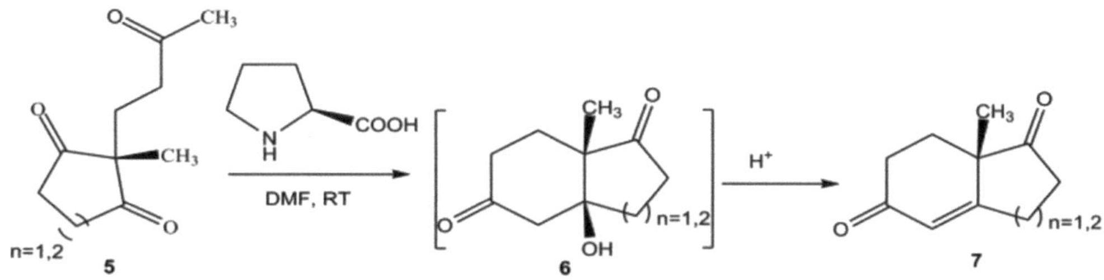

Scheme 1. L-Proline organocatalyzed intramolecular aldol condensation reaction.

A variety of small organic molecules have since been employed as asymmetric organocatalysts such as proline [13], proline derivatives [14], cinchona derivatives [15], binapthol derivatives [16] (Marouka's catalyst), and guanidinium-based catalysts [17] in various chemical reactions. Moreover, lately sincere efforts have been made to design and synthesize newer organocatalysts having superior and effective properties in asymmetric organic transformation reactions. Here, we provide an overview on the recent developments in the field of guanidine-based catalysts and their ability to act as chiral catalysts in various chemical reactions.

Generally, organocatalysts can be subdivided into various categories based on their binding ability with the substrate through covalent bond, noncovalent interactions such as hydrogen bonding or electrostatic/ion pair interactions as shown in **Figure 2**.

In the first category, chiral organocatalyst forms covalent bond with an achiral substrate leading to a chiral transition complex including enamine and iminium activation. For such type of catalytic activation, proline and proline-derived secondary amines or cinchona alkaloid-derived primary amines have been widely used as asymmetric organocatalyst in many organic reactions. *N*-Heterocyclic carbene and phosphine-derived organocatalysts for stereoselective Morita-Baylis-Hillman or aza-Morita-Baylis-Hillman type reactions using covalent bond forming activation are well documented [18]. The second category includes asymmetric organocatalyst that forms H-bond with an achiral substrate resulting in a chiral transition state. The example includes Brønsted acid catalysts such as derived phosphoric acids, phosphoryl triflylamides, thiourea, and squaramide which form H-bond with the substrate through activation [19]. The last category includes electrostatic interactions between chiral organocatalyst and achiral substrate for an active chiral transition state. List et al. developed newer concept involving asymmetric counteranion-directed catalysis (ACDC) for the activation mode where a chiral counteranion formed an enantioselective product through an activated transition state [20]. Here, chiral ammonium salts or other protonated bifunctional catalysts such as thioamide or guanidinium catalysts were used as chiral countercations and these are also called phase-transfer catalysts.

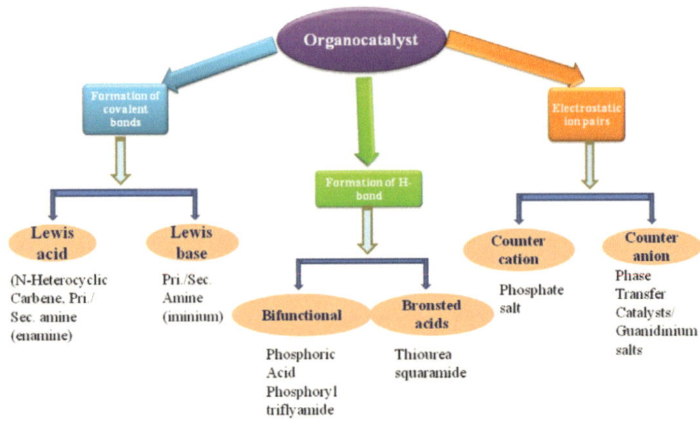

Figure 2. General classification of organocatalysts based on their activation mode.

2. Guanidine-based asymmetric organocatalysis

Guanidine, discovered over 150 years ago, is well recognized as a very strong base (superbase). Useful chemical functionalities are shown by guanidine and their corresponding salts. Free guanidine displays dual behavior, Brønsted basicity, as well as hydrogen bond donating and accepting abilities [21]. While guanidinium salts show weak Brønsted acidity, cationic hydrogen bond donating capability and the possibility of delocalizing guanidinium cationic π-Lewis acids are shown in **Figure 3**.

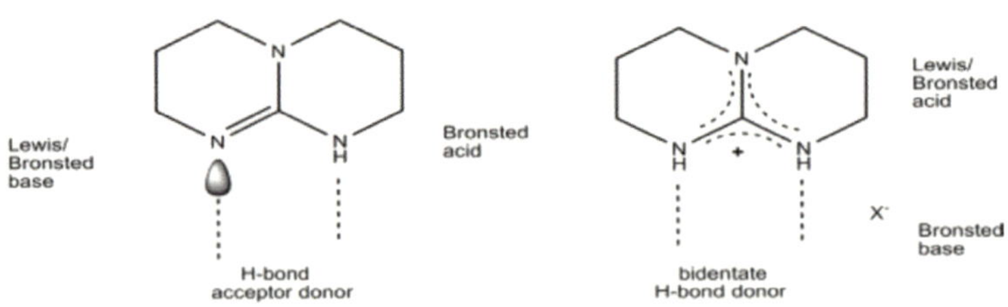

Figure 3. Guanidine as a Brønsted base as well as a hydrogen bond donor/acceptor.

http://dx.doi.org/10.5772/63268

Despite the diverse functionalities present in the guanidine group, its synthesis for newer organocatalytic applications of guanidine is a relatively new research area in chiral compound synthesis. Structurally, guanidine organocatalysts can be classified into several categories such as open chain **8**, monocyclic **9**, and bicyclic **10** structures as shown in **Figure 4**.

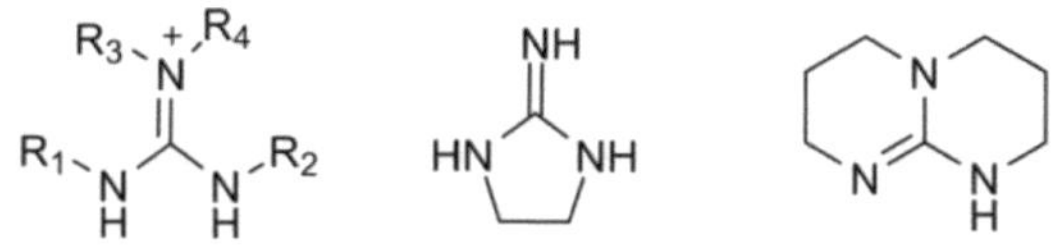

8 9 10

Figure 4. Structures of bicyclic, monocyclic, and open-chain guanidine scaffolds.

The guanidine and guanidinium salts possess similar features as urea and thiourea, such as dual hydrogen bonding, which is a key interaction in the electrophilic activation as well as transition state organization. Najera et al. for the first time used open chain guanidine **11** in the early 1990s for enantioselective organocatalyzed Henry reaction [22], affording amino alcohols up to 54% ee (**Scheme 2**).

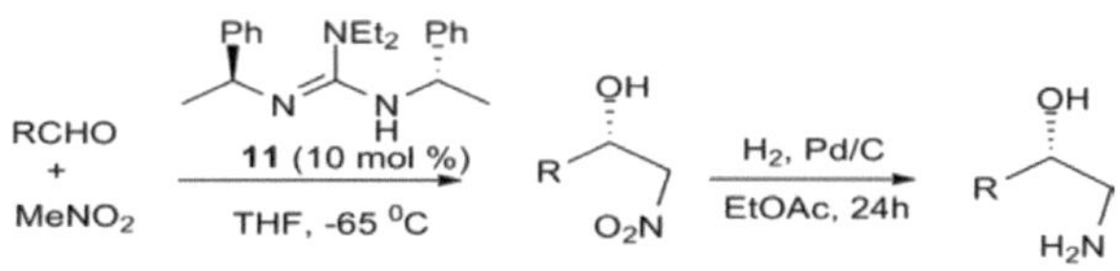

Scheme 2. Open chain guanidinium **11** organocatalyzed Henry reaction.

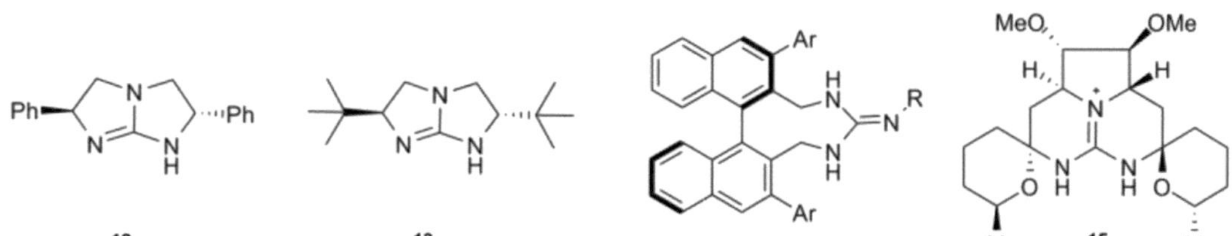

12 13 14 15

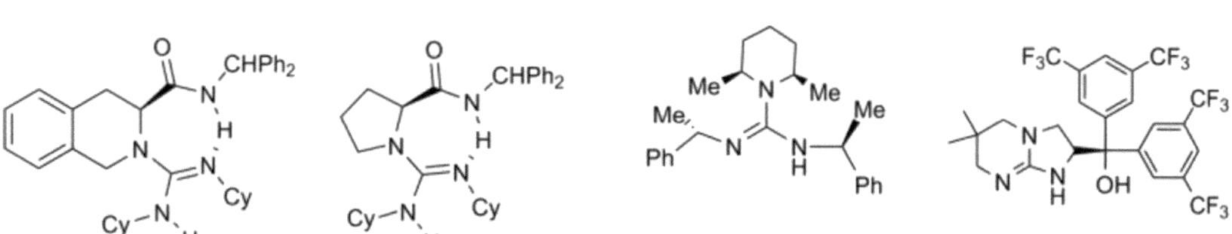

16 17 18 19

Figure 5. Guanidine-containing asymmetric organocatalysts.

Later in 1999, Lipton et al. reported cyclic dipeptide organocatalyzed highly enantioselective Strecker reaction using 10 mol% of organocatalyst [23]. Further, Corey and Grogan repeated the Strecker reaction using bicyclic guanidinium chiral organocatalyst [24]. Nagasawa et al. used pentacyclic guanidine organocatalyst **15** for the enantioselective alkylation reactions [25]. Tan et al. reported bicyclic guanidine to catalyze stereoselective alkylation and Michael reactions [26]. Subsequently, several chiral guanidinium organocatalyst were used for stereoselective organic transformation reactions.

Some of the well-documented guanidine-based asymmetric organocatalysts are shown in **Figure 5**.

3. Applications of asymmetric organocatalysts

Asymmetric organocatalysis is recognized as an independent synthetic toolbox in addition to asymmetric metallic and enzymatic catalysis for the synthesis of chiral molecules (**Figure 6**).

Figure 6. Application of asymmetric organocatalysis in various stereoselective organic transformation reactions.

In the present chapter, we have focused on the enantioselective reactions catalyzed by chiral guanidinium and their salts.

3.1. Asymmetric alkylation

Asymmetric alkylation by phase-transfer catalyst is a well-established approach. Nagasawa et al. designed pentacyclic guanidinium salts for the enantioselective alkylation of glycinate Schiff base [25]. Glycinate Schiff base underwent alkylation reaction with various alkyl halides under phase-transfer conditions in the presence of guanidinium salt **15** (**Table 1**). It was proposed that the spiro ether rings determined the configuration of the newly formed chiral center. The methyl group of the spiro ether ring played a crucial role in the enantioselectivity.

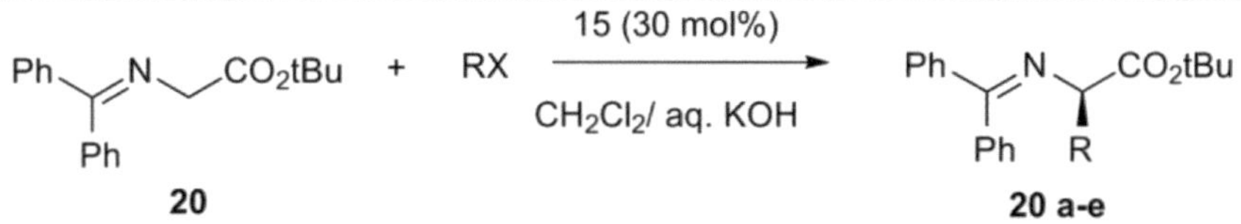

Entry	RX	Product	*t* (h)	Yield (%)	ee (%)
1	BnBr	20a	160	55	90
2	MeI	20b	145	80	76
3	OctI	20c	145	83	80
4	CH_2=$CHCH_2Br$	20d	140	61	81
5	CH_2=$C(Me)CH_2Br$	20e	145	85	81

ee = enantiomeric excess.

Table 1. Pentacyclic guanidinium **15** catalyzed asymmetric alkylation reaction with various alkyl halides.

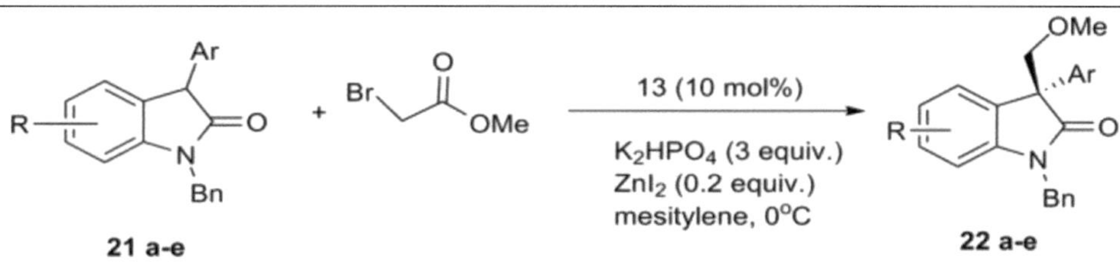

Entry	*t* (h)	21 (*R*, Ar)	Yield (%)	ee (%)
1	44	21a (5-F, Ph)	98	94
2	37	21b (5-Cl, Ph)	99	94
3	51	21c (5-Br, Ph)	99	93
4	59	21d (5-Me, Ph)	96	93
5	63	21e (MeO, Ph)	90	93

ee = enantiomeric excess.

Table 2. Alkylation of 3-substituted 2-oxiindoles using activated bromomethanes.

Tan et al. used bicyclic guanidinium **13** as chiral phase-transfer catalyst [27] for enantioselective alkylation of 3-substituted 2-oxiindoles shown in **Table 2**.

The position and electronic properties of the substituents on the para- and meta-positions of aromatic ring at the C-3 position of 3-aryl-2-oxindoles did not affect the enantioselectivity.

Guanidine containing *s*-triazene asymmetric organocatalyst has been synthesized and used for asymmetric alkylations [28] as shown in **Table 3**.

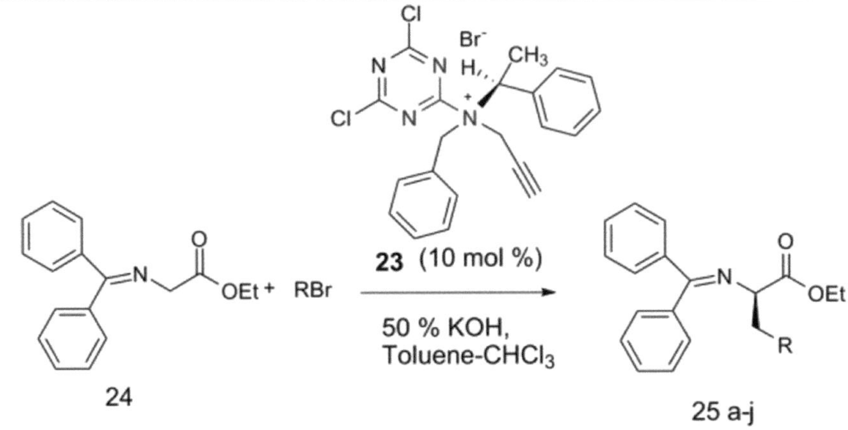

Entry	Electrophile (RBr)	Product	*t* (h)	Yield (%)	ee (%)
1	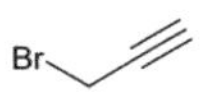	25a	3.5	75	95
2	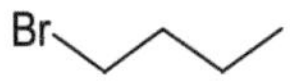	25b	3.5	85	89
3	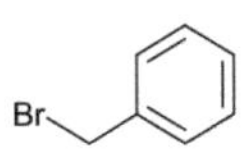	25c	4.5	90	82
4	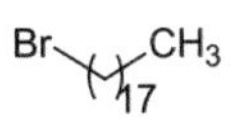	25d	6.0	67	81
5	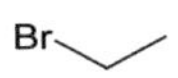	25e	4.5	55	75
6	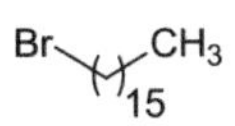	25f	5.5	85	85
7	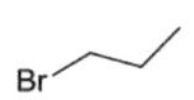	25g	5.5	80	77
8	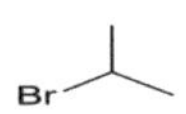	25h	4.5	76	68
9	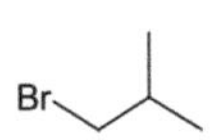	25i	4.5	90	83
10	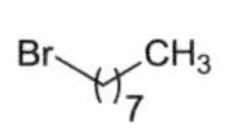	25j	4.5	85	90

ee = enantiomeric excess.

Table 3. Alkylation of glycinate Schiff base using s-triazene based asymmetric organocatalyst.

3.2. Asymmetric aldol reaction

List et al. reported enantioselective intermolecular aldol reaction using proline organocatalyst [17]. In these years, hundreds of research articles were published on the stereoselective

http://dx.doi.org/10.5772/63268

aldol reactions using various asymmetric organocatalysts. Moreover, *l*-proline and guanidine salts as cocatalysts were also used for enantioselective aldol reaction [29] using various aldehydes 42 a-k, which reacted with chloroacetone (**Table 4**, Entries 1–5) smoothly, with good yield and high regio-, diastereo-, and enantioselectivity.

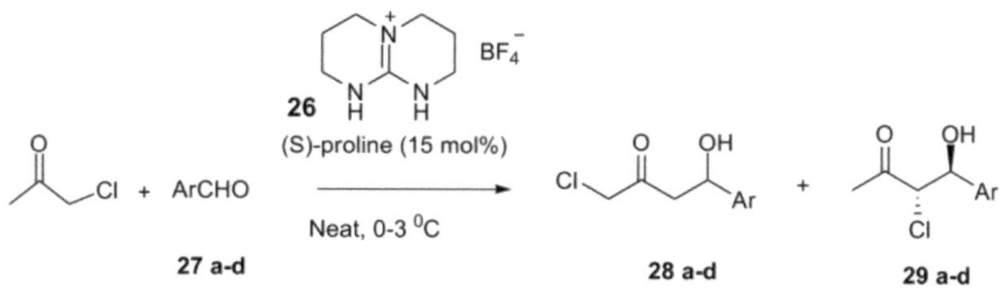

Entry	ArCHO		Conversion	Regioselectivity	d.r. (%)	ee (%)
1	27a	4-NO_2–C_6H_4	99	96:4	91:9	98
2	27a	4-NO_2–C_6H_4	99	99:1	83:17	95
3	27b	3-NO_2–C_6H_4	97	96:4	92:8	97
4	27c	2-NO_2–C_6H_4	98	>99:1	93:7	97
5	27d	4-CO_2Me–C_6H_4	96	99:1	91:9	97

[a]d.r. = distereomeric ratio; ee = enantiomeric excess.

Table 4. Synthesis of chlorohydrins using (S)-proline/guanidine salt as cocatalyst.

3.3. Asymmetric epoxidation

Taylor et al. investigated the asymmetric epoxidation [30] reaction using chiral guanidines 30 a–c (**Figure 7**).

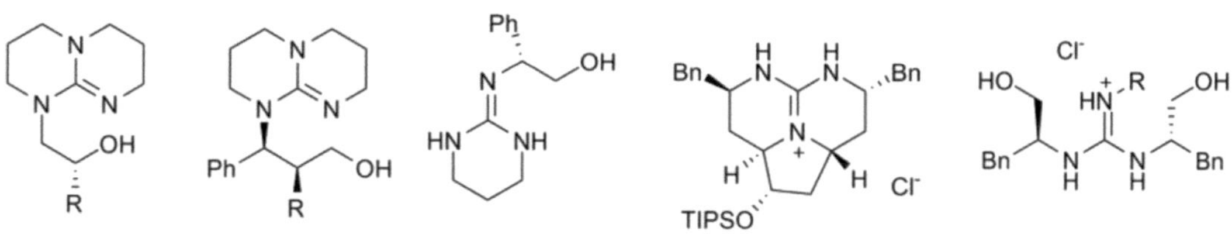

30 a : R= Ph
30 b : R= 3-ClC_6H_4 **30 c** **31** **32** **33** R= $C_{18}H_{37}$

Figure 7. Structure of guanidine catalyst for asymmetric epoxidation.

Epoxide of amidoquinone analogs were obtained in poor-to-moderate ee using the stoichiometric amounts of these guanidine-based chiral organocatalysts (**Scheme 3**).

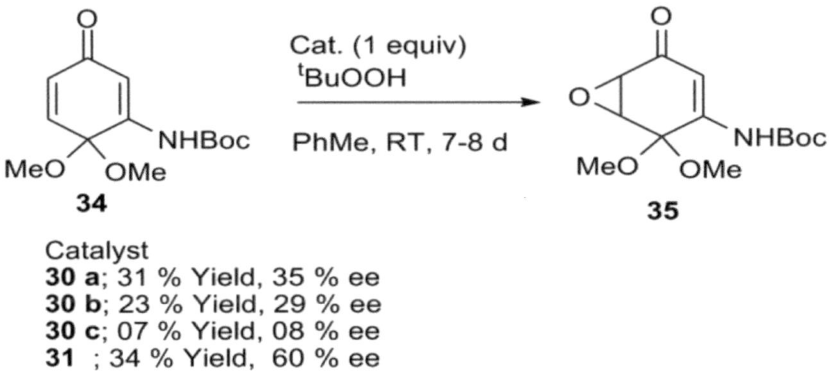

Scheme 3. Asymmetric epoxidation of amidoquinone using various chiral guanidine organocatalysts.

Acyclic guanidine **31** when used in stoichiometric amount shows better ee (60%) in comparison to cyclic guanidines (30a–c). Epoxidation of trans-chalcone and 2-methyl-napthoquinone with chiral guanidine was also carried out. Good yields were obtained but the ee were very poor [31].

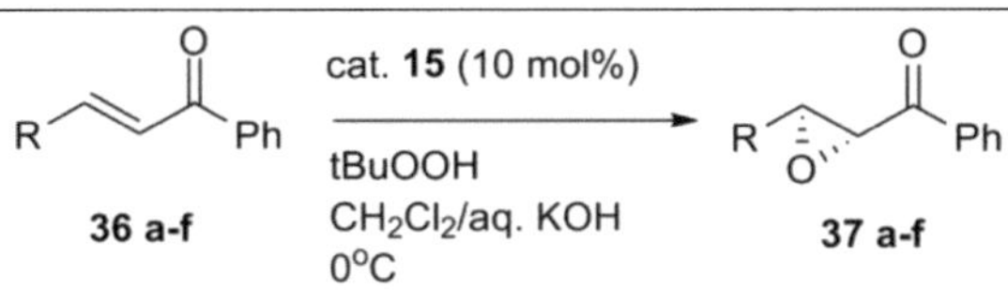

Entry	*R*	*t* (h)	Yield (%)	ee (%)
1	Ph	110	35	39
2	2-naphthyl	140	51	50
3	1-naphthyl	140	77	60
4	9-anthracenyl	140	>99	35
5	4-NO_2 C_6H_4	130	82	38
6	4-$MeOC_6H_4$	160	22	36

ee = enantiomeric excess.

Table 5. Asymmetric epoxidation reactions of various chalcones with pentacyclic guanidinium salt 15.

Nagasawa et al. used pentacyclic guanidine salt **15** as phase-transfer catalyst for the epoxidation of chalcones [32] up to 60% ee ratio (**Table 5**).

3.4. Asymmetric Diels-Alder reaction

Enantioselective cycloaddition is a large area of research catalyzed by Lewis acids [33]. However, the base catalyzed stereoselective Diels-Alder reaction has remained largely unexplored. Ma et al. used chiral guanidines **38** as asymmetric organocatalyst for the cycloaddition reaction of anthrone and methyl maleimide (**Scheme 4**) [34].

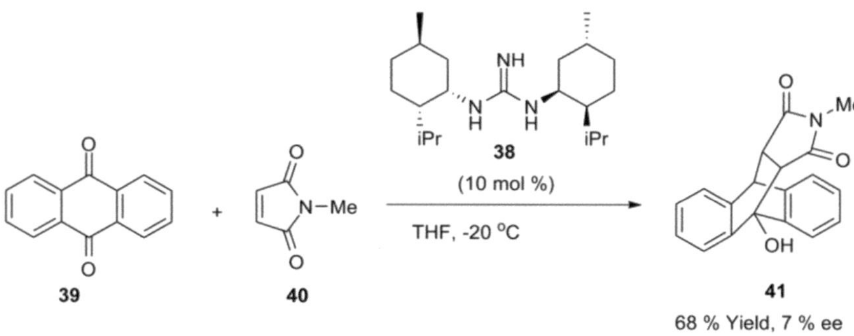

Scheme 4. Diels-Alder reaction between anthrone and methyl maleimide.

Tan et al. describe a highly enantioselective guanidine catalyzed Diels-Alder reaction between anthrones and activated olefins [35].

3.5. Asymmetric Friedel-Craft reaction

Friedel-Craft alkylation has been widely used for the synthesis of relevant and promising biological entities [36]. Despite the aromatic substitution reactions, catalytic and asymmetric versions of Friedel-Craft reactions have been described in the mid 1980s. Recently, chiral organocatalysts such as imidazolidinone, cinchona alkaloids, diaryl prolinol derivatives, phosphoric acids, thiourea-mediated and guanidine-based catalysts have become more popular for these transformations.

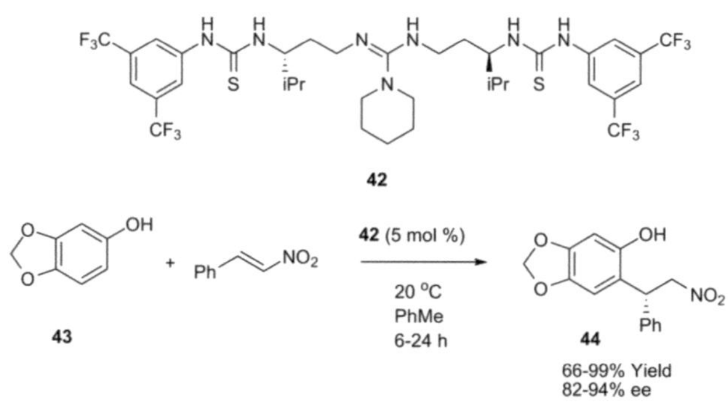

Scheme 5. Chiral organocatalyzed 1,4-type Friedel-Craft reaction of phenols.

Nagasawa et al. present conformationally flexible stereoselective guanidine/bisthiourea organocatalysts for chemo-, regio-, and enantioselective 1,4-type Friedel-Craft reaction of phenols as shown in **Scheme 5** [37].

3.6. Asymmetric Henry reaction

The Henry reaction (nitro-aldol) is one of the oldest C–C bond formation reactions in organic synthesis. Shibasaki et al. in 1992 for the first time reported the asymmetric version of the Henry reaction [38]. Later, Najera et al. in 1994 used guanidine organocatalyst for the enantioselective Henry reaction [22]. Since then, various newer guanidine-based chiral organocatalysts for the asymmetric Henry reaction have been developed. Some bifunctional acyclic/cyclic and bisguanidine catalysts were also designed for stereoselective Henry reaction (**Figure 5**).

Nagasawa et al. used effective linear guanidine-thiourea-based bifunctional catalyst 14 for an enantio- as well as diastereoselective Henry reaction [39]. Chiral guanidine-amide organocatalyst **17** has been used for an efficient asymmetric aza-Henry reaction [40] of isatin-derived ketimines (**Scheme 6**).

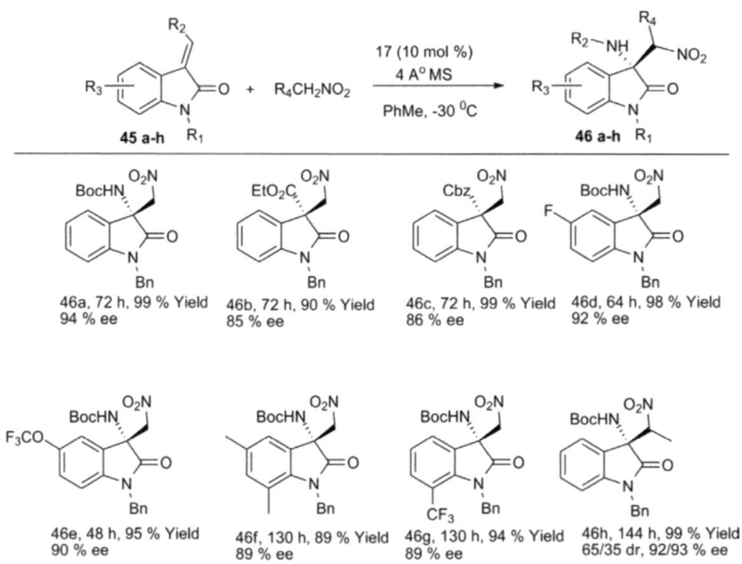

Scheme 6. Organocatalyst **17** catalyze aza-Henry reaction.

Ma et al. studied the diastereoselective Henry reactions [41] of α-aminoaldehyde with nitromethane using open-chain chiral guanidine organocatalyst.

Murphy and coworkers used tetracyclic guanidinium salt for the Henry reaction of nitromethane and isovaleraldehyde with 20% enantioselectivity [42].

Recently, Zhao and coworkers developed a new protocol for the highly stereoselective tandem Henry-Michael reaction using Misaki-Sugimura guanidine catalyst **19** to synthesize trisubstituted cyclohexanols [43]. Using the optimized reaction conditions, the desired trisubstituted cyclohexanols obtained in both high enantioselectivities and diastereoselectivities are shown in **Table 6**.

http://dx.doi.org/10.5772/63268

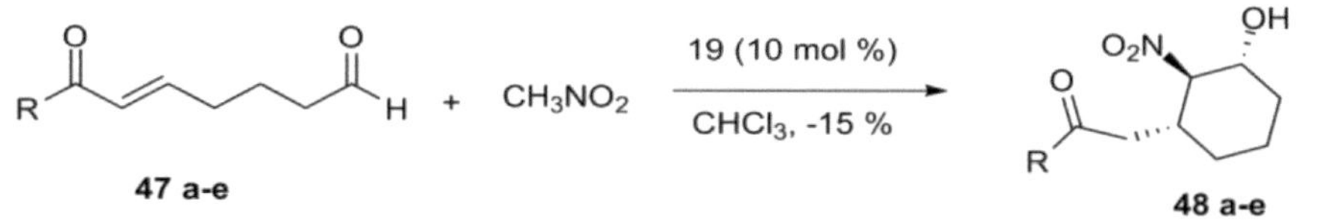

Entry	R	Product	Yield (%)	d.r. (%)	ee (%)
1	Ph	48a	99	>99:1	98
2	4-FC_6H_4	48b	98	>99:1	96
3	4-ClC_6H_4	48c	95	>99:1	96
4	4-BrC_6H_4	48d	99	>99:1	97
5	4-CNC_6H_4	48e	98	>99:1	98

d.r.= diastereomeric ratio; ee = enantiomeric excess.

Table 6. Misaki-Sugimura guanidine 19 catalyzed tandem Henry-Michael reaction.

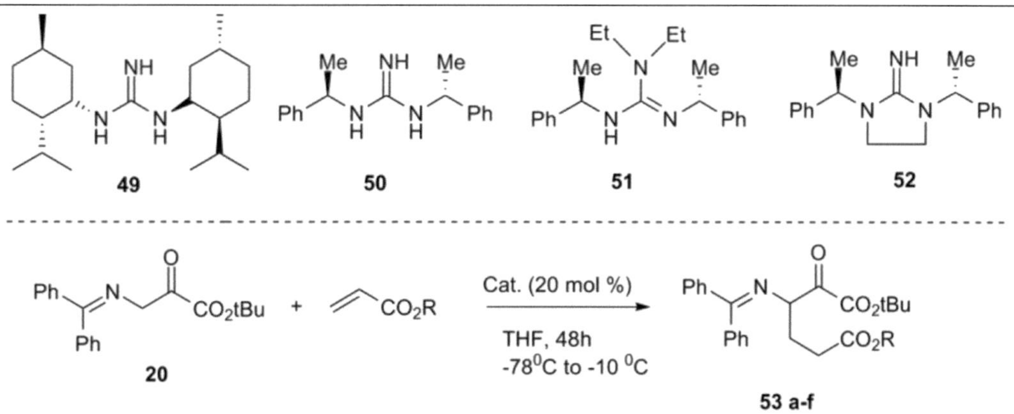

Entry	Catalyst	R	Product	Yield (%)	ee (%)
1	49	Et	53a	99	30
2	50	Et	53b	95	6
3	52	Et	53c	97	17
4	51	Et	53d	85	26
5	49	Me	53e	95	16
6	49	tBu	53f	98	30

ee = enantiomeric excess.

Table 7. Asymmetric conjugate additions of glycinate to various acrylates using various organocatalysts.

3.7. Asymmetric Michael reaction

Michael reaction is the most common approach toward C–C or C–X bond formation via conjugate addition of nucleophiles to electron-deficient alkenes [44]. Michael reaction between glycinate and ethyl acrylate was described by Ma et al. using chiral guanidine organocatalysts (**Table 7**) [34a].

Chiral guanidine catalyst **49** was also used for the enantioselective synthesis of the core structure of (−)-huperzine **A** by the Michael-aldol annulations reaction [34b]. Ishikawa et al. catalyzed the enantioselective Michael reaction of glycinate with various Michael acceptors using the chiral guanidine organocatalyst **31** under solvent-free condition affording excellent enantiomeric excess (**Scheme 7**) [45].

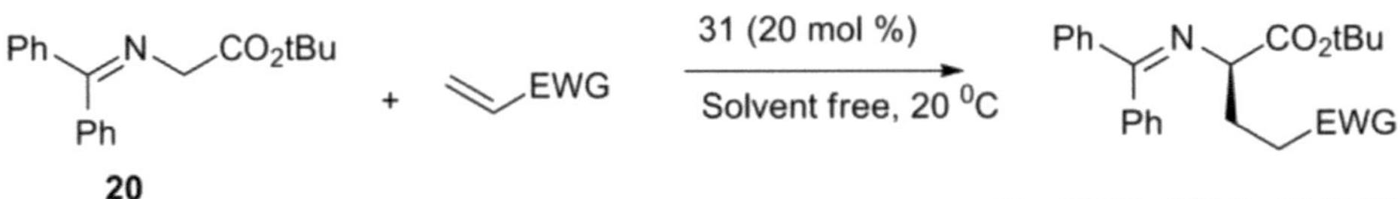

54a, EWG= CO_2Et, 85 % Yield, 97 % ee

54b, EWG= CO_2Me, 98 % Yield, 93 % ee

54c, EWG= CO_2CN, 79% Yield, 55 % ee

Scheme 7. Enantioselective conjugate addition reaction of glycinate with various Michael acceptors.

Ishikawa et al. also attempted the Michael reaction [46] of 2-cyclo-penten-1-one with dibenzyl malonate using the same guanidine organocatalyst **31** (**Scheme 8**).

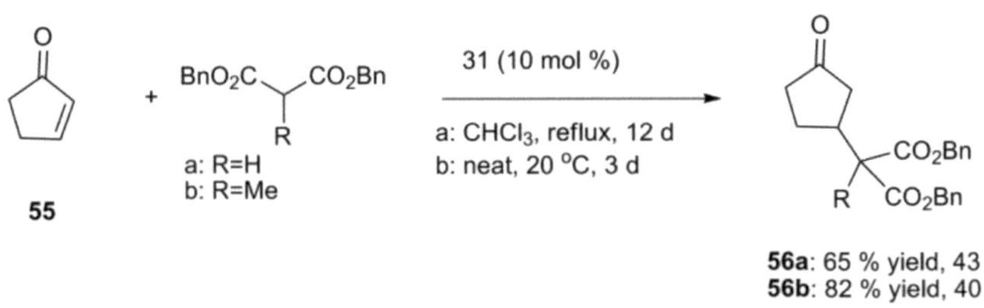

56a: 65 % yield, 43 % ee
56b: 82 % yield, 40 % ee

Scheme 8. Michael reactions between cyclo-penten-1-one and dibenzyl malonate using the guanidine organocatalyst **31**.

Terada et al. developed axially chiral guanidine organocatalyst **14** that facilitates the highly enantioselective 1,4-Michael reaction with 1,3-dicarbonyl compounds [47] as shown in

Table 8. Michael conjugated products provide various types of optically active nitroalkane derivatives of synthetic and biological importance.

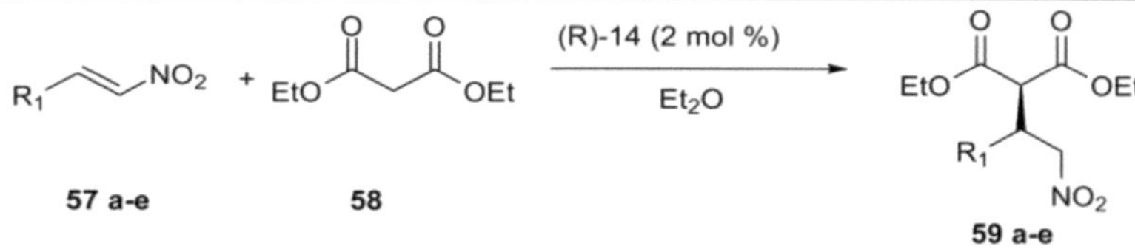

Entry	R_1	Product	*t* (h)	Yield (%)	ee (%)
1	2-$MeOC_6H_4$–	59a	2	98	97
2	2-BrC_6H_4–	59b	2	>99	98
3	2-$NO_2C_6H_4$–	59c	4	86	91
4	3-$MeOC_6H_4$–	59d	2	94	94
5	3-BrC_6H_4–	59e	2	90	95

ee = enantiomeric excess.

Table 8. 1,4-Michael reaction of various conjugate reactions with diethyl malonate using organocatalyst **14**.

Linton et al. designed pentapeptide organocatalyst incorporated with arginine for the Michael reaction of nitrocarbonyl compounds [48]. Tan et al. used guanidine organocatalyst **12** for the Michael reaction between anthrone **60** and maleimides **61 (a–d)** to obtained cycloadducts **62 (a–d)** in excellent yields and enantioselectivities (**Table 9**) [35a, 49].

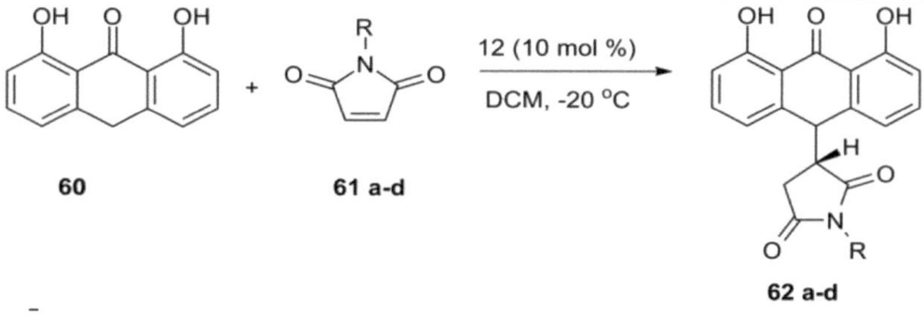

Entry	R	Product	*t* (h)	Yield (%)	ee (%)
1	Ph	62 a	7	80	99
2	2-$NO_2C_6H_4$	62 b	8	87	97
3	3,4-$Cl_2C_6H_3$	62 c	8	89	98
4	Bn	62 d	8	86	93

ee = enantiomeric excess.

Table 9. Bicyclic guanidine organocatalyst 12 catalyzed enantioselective Michael reaction.

Bicyclic guanidine organocatalyst **12** worked well with maleimide (**Table 9**) and other activated olefins (**Table 10**) with excellent enantioselectivities and regioselectivities.

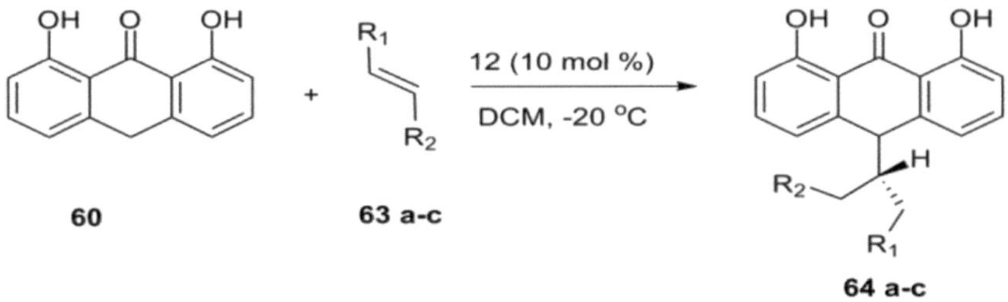

Entry	R_1	R_2	Product	*t* (h)	Yield (%)	ee (%)
1	CO_2Me	PhCO	64 a	6	92	98
2	CO_2Et	MeCO	64 b	7	92	95
3	CN	CN	64 c	7	90	94

ee = enantiomeric excess.

Table 10. Asymmetric Michael reaction of anthrone and activated olefins using chiral organocatalyst **12**.

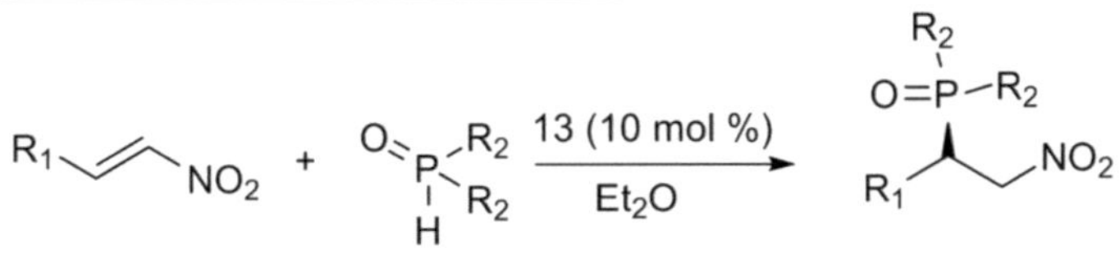

Entry	R_1	R_2	*T* (°C)	*t* (h)	Yield (%)	ee (%)[a]
1	Ph	Ph	0	13	64	60 (96)
2	Ph	4-FC_6H_4	0	14	92	60 (>99)
3	Ph	4-PhC_6H_4	0	40	85	50 (91)
4	Ph	2-EtC_6H_4	−40	40	77	75
5	Ph	2-naphthyl	0	6	92	65 (99)

ee = enantiomeric excess.
[a]ee values obtained after single recrystallization.

Table 11. Phospha Michael reaction of various diaryl phosphine oxide and nitroalkenes using bicyclic guanidine organocatalyst 13.

Tan et al. used chiral bicyclic guanidine organocatalyst for the phospha-Michael reaction of nitroalkenes (**Table 11**) [50]. Various nitroalkenes with di-(1-naphthyl) phosphine oxide at −40°C gave excellent enantioselectivities.

Terada et al. demonstrated that axially chiral binaphthyl organocatalyst **14** (Ar = 3,5-di-tBuC_6H_4; R = Bn) can used for the phospha-Michael reaction of diphenyl phosphite to nitroalkenes with high enantioselectivities [51].

Ishikawa et al. investigated the 6-exo-trig intramolecular oxa-Michael cyclization reaction for the chiral chromane **67** synthesis using guanidine organocatalyst [52] **65** as per **Scheme 9**. The E/Z geometry of the α, β unsaturated ester played a crucial role in the enantiomeric excess determination of chromane moiety.

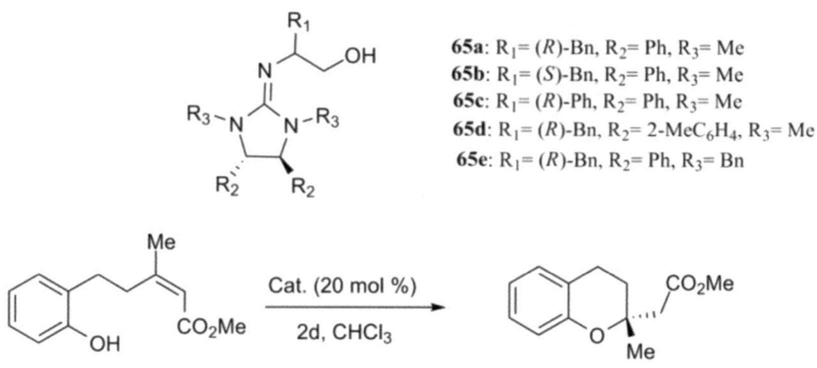

Scheme 9. Intramolecular oxa-Michael cyclization reaction of the chiral chromane synthesis.

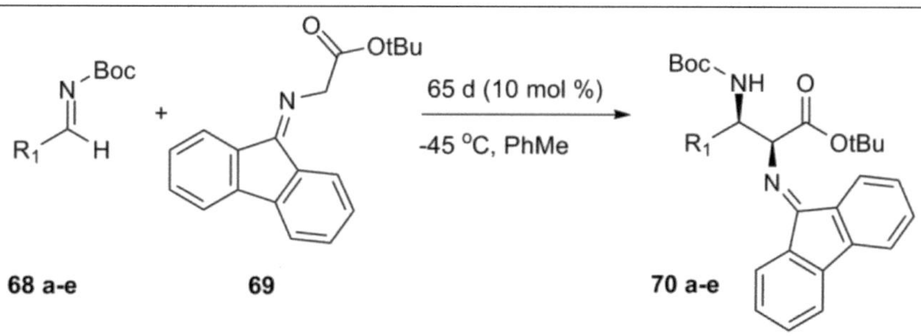

Entry	R_1	Product	*t* (h)	Yield (%)	Syn:Anti	ee (%)
1	Ph	70 a	12	>99	>99:1	96
2	4-MeOC_6H_4	70 b	48	76	36:1	90
3	2-furyl	70 c	36	88	9:1	98
4	$Ph(CH_2)_2$	70 d	24	87	29:1	92
5	*c*-C_6H_{11}	70 e	48	84	11:1	96

ee = enantiomeric excess.

Table 12. Enantioselective Mannich reactions of various *N*-Boc protected imines using guanidine organocatalyst 65 d.

3.8. Asymmetric Mannich reaction

The asymmetric Mannich reaction ranks among the most potent enantioselective and diastereoselective C–C bond forming reactions to obtain chiral β-aminocarbonyl compounds from imines. Asymmetric organocatalytic reactions had been successfully developed for the well-known Mannich reaction in particular. In the Mannich reaction, a key species, an iminium intermediate is formed which is susceptible to nucleophilic attack. Recently, Kobayashi et al. [53] reported the Mannich reaction of fluorenone imine of glycine ester and its phosphonic acid analogs using the guanidine organocatalyst **65d** (**Table 12**).

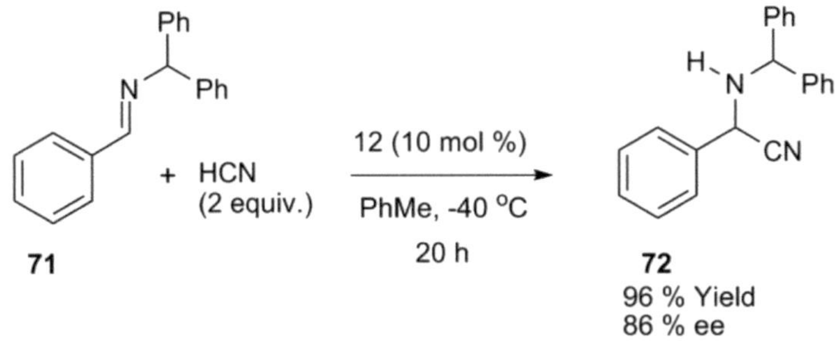

Scheme 10. Strecker reaction using bicyclic guanidine organocatalyst **12**.

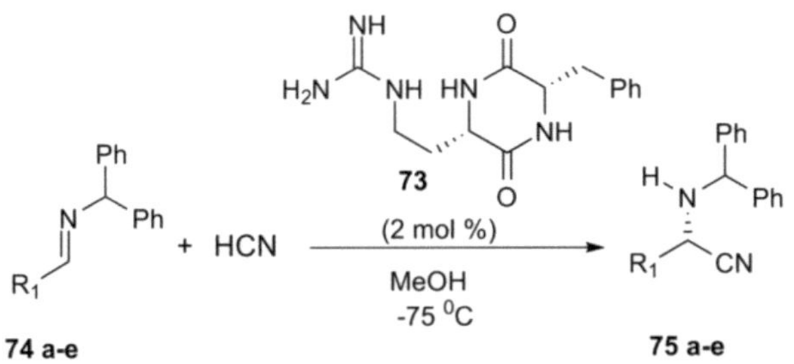

Entry	R_1	Product	Yield (%)	ee (%)
1	Ph	75 a	97	>99
2	4-ClC_6H_4	75 b	94	>99
3	4-$MeOC_6H_4$	75 c	90	96
4	3-ClC_6H_4	75 d	80	>99
5	3-$MeOC_6H_4$	75 e	82	80

ee = enantiomeric excess.

Table 13. Asymmetric Strecker of *N*-benzhydryl imines catalyzed by dipeptide organocatalyst **73**.

3.9. Asymmetric Strecker reaction

Strecker reaction is an excellent way for the synthesis of α-amino acids [54]. Lipton group in 1996 for the first time reported the asymmetric version of the Strecker reaction [23]. In addition, the metal-catalyzed asymmetric cynation and chiral organocatalytic process had been used for the enantioselective Strecker reaction. Interestingly, chiral organocatalyst possess high catalytic properties for the hydrocynation reaction. Corey group used chiral bicyclic guanidine as an efficient catalyst in the asymmetric addition of hydrogen cyanide to imine [24]. The hydrocynation of the benzaldehyde-derived imine gave the corresponding (*R*)-amino nitrile in 96% yield and high enantiomeric excess (86%) using 10 mol% of bicyclic guanidine organocatalyst **12** shown in **Scheme 10**.

Lipton et al. used guanidine-based dipeptide organocatalyst **73** for the stereoselective Strecker reaction (**Table 13**) [23]. Further, the replacement of the guanidine functional group of the organocatalyst **73** with an imidazole ring failed to achieve any enantioselectivity. It confirms that the guanidine group plays a crucial role in enantioselectivity determination.

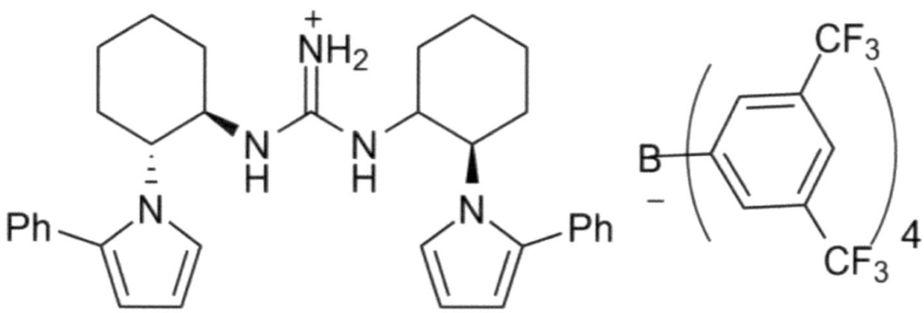

76

Figure 8. C_2-Symmetric guanidinium salt **76** for Claisen rearrangement.

3.10. Asymmetric Claisen rearrangement

Rainer Ludwig Claisen discovered [3, 3]-sigmatropic rearrangement of allyl vinyl ethers which led to one of the most powerful C–C bond forming reactions [55]. Jacobsen and coworkers used catalytic amount of the C_2-symmetric guanidinium salt **76** for the asymmetric Claisen rearrangement (**Figure 8**) [56] and obtained greater enantioselectivity with high yield (**Table 14**).

3.11. Asymmetric reduction reaction

Basavaih et al. reported the borane-mediated asymmetric reduction of phenacyl bromide using the chiral guanidine organocatalyst **79** [57]. When the reaction was carried out at room temperature, it gave *R*-configured alcohol with 37% ee, while under reflux condition, the *S* alcohol was obtained with improvement in ee value to 83% (**Scheme 11**).

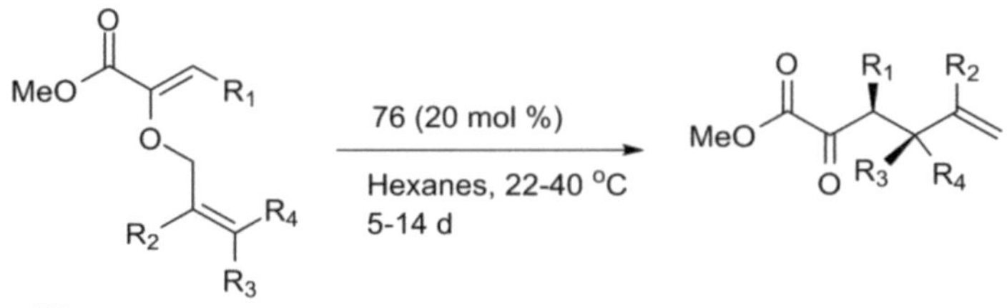

Entry	R_1	R_2	R_3	R_4	Product	Yield (%)	ee (%)
1	Me	H	H	H	78 a	80	92
2	Et	H	H	H	78 b	86	92
3	Et	H	H	*n*Pr	78 c	92	85
4	Et	H	H	Ph	78 d	91	81
5	Me	Me	H	H	78 e	73	96

ee = enantiomeric excess.

Table 14. Guanidine salt **76** catalyzed enantioselective Claisen rearrangement.

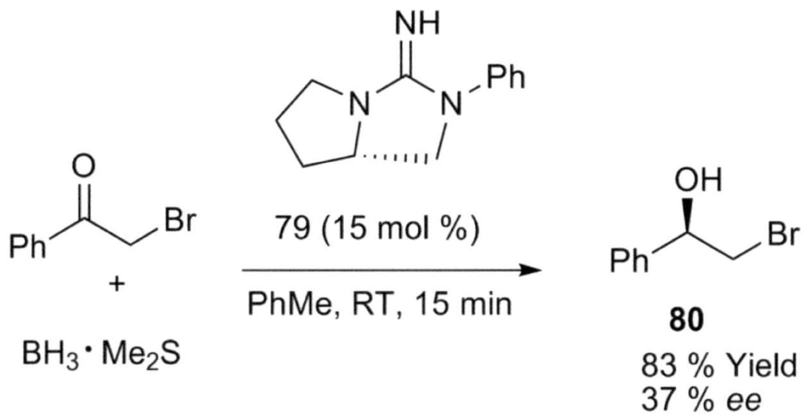

Scheme 11. Asymmetric reduction of phenylacyl bromide using chiral guanidine organocatalyst **79**.

3.12. Asymmetric amination reaction

Asymmetric electrophilic amination reaction of 1,3-dicarbonyl compounds was achieved by Terada et al. with a C_2-symmetrical axially chiral guanidine organocatalyst **81** which has a seven-membered ring structure [58]. The reaction was conducted using 2 mol% catalyst loading with di-*tert*-butyl azidocarboxylate at –60°C temperature. Bulkiness of the azidocarboxylate played a crucial role in the enantioselectivity. The scope of the reaction is shown in **Table 15** with the optimal reaction conditions.

http://dx.doi.org/10.5772/63268

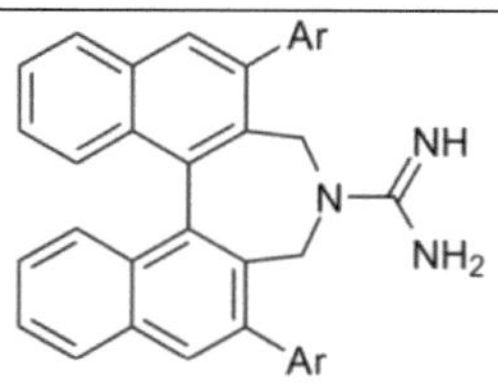

81, Ar= 4-(3,5-di-*t*BuC$_6$H$_3$)C$_6$H$_4$

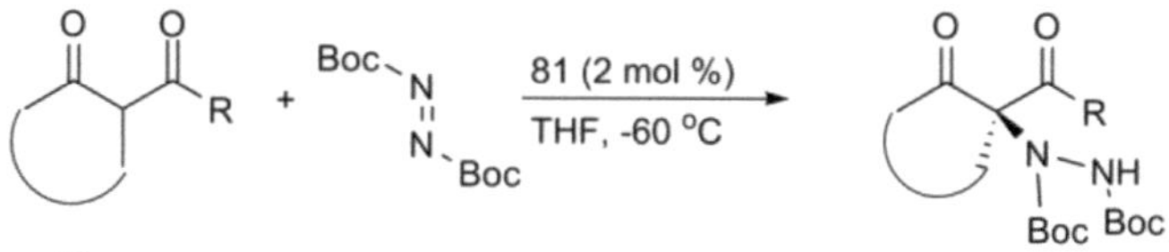

82 a-e

83 a-e

Entry	Reactant	R	Product	*t* (h)	Yield (%)	ee (%)
1	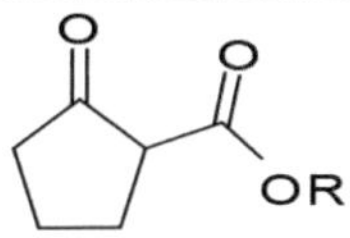	–CH_2CH_3	83 a	4	>99	97
2	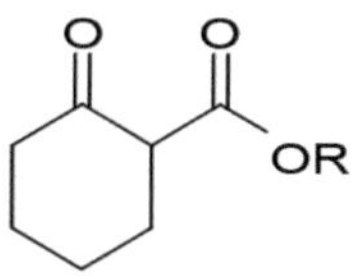	–CH_2CH_3	83 b	24	>99	98
3	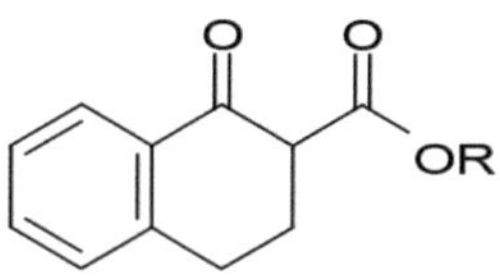	–CH_2CH_3	83 c	1	>99	97
4	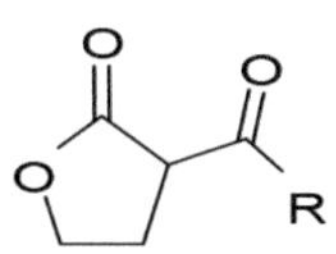	–CH_3	83 d	24	>99	15
5	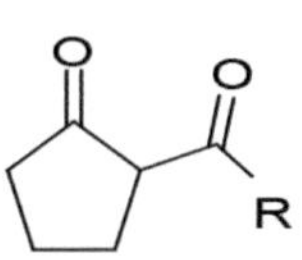	–CH_3	83 e	5	99	91

ee = enantiomeric excess.

Table 15. Asymmetric electrophilic amination reactions of various 1,3-dicarbonyl ketones.

4. Other important reactions using chiral guanidine organocatalyst

4.1. Asymmetric protonation reaction

Protonation reaction is a direct approach for the preparation of carbonyl compounds with a stereogenic center of enolates. A transient enolate is first generated through a conjugate addition reaction, followed by an in situ enantioselective protonation reaction. Tan et al. investigated the protonation of 2-phthalimidoacrylate **84** with thiophenols and obtained a series of arylthiols **85** with excellent yields and enantioselectivities (**Table 16**) [59].

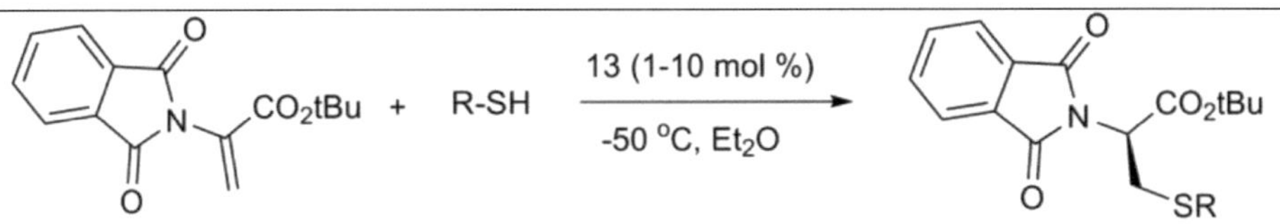

84 **85 a-e**

Entry	R	20 (mol %)	Product	*t* (h)	Yield	ee (%)
1	Ph	10	85 a	0.5	99	90
2	2-$CF_3C_6H_4$	10	85 b	3	99	93
3	2-$MeO_2CC_6H_4$	10	85 c	2.5	98	90
4	4-BrC_6H_4	10	85 d	4	99	90
5	4-*t*BuC_6H_4	10	85 e	3	92	93

ee = enantiomeric excess.

Table 16. Enantioselective protonations of 2-phthalimidoacrylate using the bicyclic guanidine organocatalyst **13**.

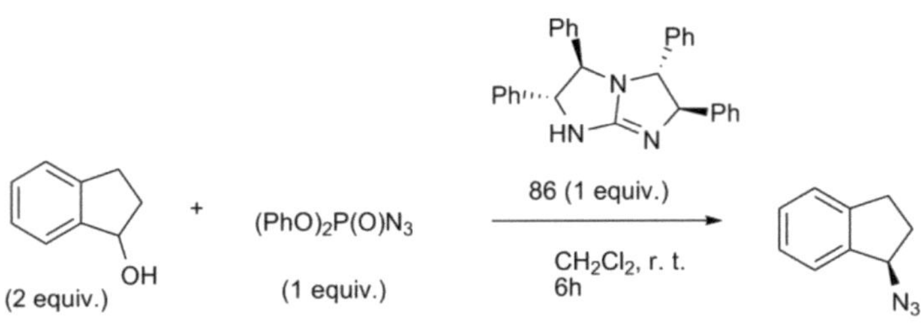

87 **88** **89**

Scheme 12. Guanidine catalyst **86** mediated asymmetric azidation reaction of 1-indanol.

4.2. Asymmetric azidation reaction

Ishikawa et al. used bicyclic guanidine catalyst for the asymmetric azidation reaction of 1-indanol in the 30% ee with diphenylphosphoryl azide (**Scheme 12**) [60]. Excess *R* isomer was produced with the use of C_2-symmetric bicyclic guanidine catalyst 86 in the 58% yield and 30% ee.

4.3. Asymmetric transamination reaction

Transamination process is a (1,3) proton-transfer reaction using imines which plays an important role in the biological systems for the production of amino acids. Berg and coworkers catalyzes transamination reaction [61] using the bicyclic guanidine organocatalyst **13** (**Table 17**).

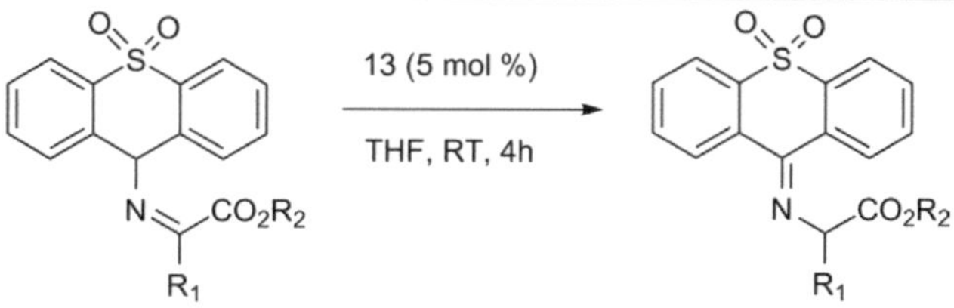

90 a-d **91 a-d**

Entry	R_1	R_2	Product	Yield (%)	ee (%)
1	Ph	Me	91 a	63	43
2	*i*Pr	Et	91 b	95	45
3	Me	Et	91 c	98	24
4	$MeO_2C(CH_2)_2$	Me	91 d	98	0

ee = enantiomeric excess.

Table 17. Stereoselective transamination reactions of an imine catalyzed by bicyclic guanidine organocatalyst **13**.

4.4. Asymmetric silylation reaction

The kinetic resolution of secondary alcohols has been done through asymmetric silylation [62] using the bicyclic guanidine organocatalyst **92** in the stoichiometric amount (**Scheme 13**).

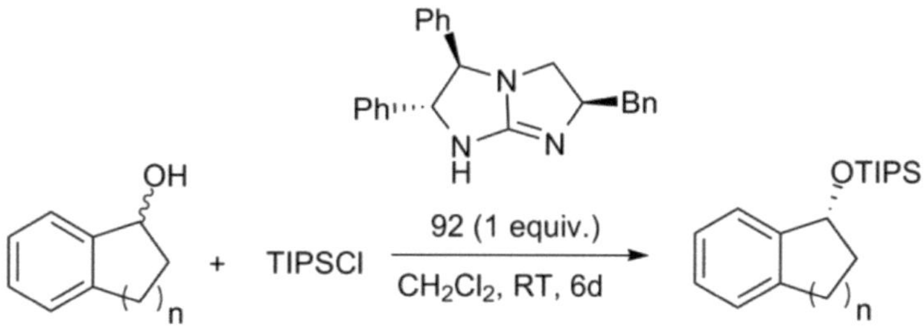

93 a-b

94 a (n=1; 79 % Yield, 58 % *ee)*

94 b (n=2; 15 % Yield, 70 % *ee)*

Scheme 13. Kinetic resolution of secondary alcohols through asymmetric silylation using chiral bicyclic guanidine organocatalyst **92**.

5. Conclusion

Guanidines containing chiral molecules have been successfully employed as chiral organocatalysts for the important asymmetric reactions. Guanidine-containing organocatalysts will continue to play an important role in asymmetric synthesis and catalysis in chemistry in coming years.

Author details

Shrawan Kumar Mangawa and Satish Kumar Awasthi*

*Address all correspondence to: satishpna@gmail.com

Chemical Biology Laboratory, Department of Chemistry, University of Delhi, Delhi, India

Appendix 1

	Reaction	Page nos.
1.	Aldol reaction	11
2.	Diels-Alder reaction	13
3.	Friedel-Craft reaction	14
4.	Henry reaction	14, 15, 16
5.	Michael reaction	16,17,18,19,20
6.	Mannich reaction	21
7.	Strecker reaction	21, 22
8.	Claisen rearrangement	23

References

[1] (a) Berkessel, A.; Groger, H. Asymmetric Organocatalysis-from Biomimetic Concepts to Powerful Methods for Asymmetric Synthesis, Wiley-VCH: Weinheim, 2005; (b) Taylor, M. S.; Jacobsen, E. N. *Angew. Chem. Int. Ed.* 2006, 45, 1520.

[2] Shao, Z.; Zhang, H. *Chem. Soc. Rev.* 2009, 38, 2745–2755.

[3] Cao, Z.-Y.; Brittain, W. D. G.; Fossey, J. S.; Zhou, F. *Catl. Sci. Technol.* 2015, 5, 3441.

[4] Seebach, D. *Angew. Chem. Int. Ed.* 1990, 29, 1320–1367.

[5] Wohler, F.; Liebig, J. *Ann. Pharm.* 1832, 3, 249–282.

[6] Liebig, J. V. *Ann. Chem.* 1860, 113, 246–247.

[7] Knoevenagel, E. *Ber. Dtsch. Chem. Ges.* 1896, 29, 172–174.

[8] Bredig, G.; Fiske, P. S. *Biochem.* 1912, 46, 7.

[9] (a) Pracejus, H.; Liebig, J. V. *Ann. Chem.* 1960, 634, 9–22; (b) Pracejus, H.; Liebig, J. V. *Ann. Chem.* 1960, 634, 23–29.

[10] Sheehan, J. C.; Hunneman, D. H. *J. Am. Chem. Soc.* 1966, 88, 3666–3667.

[11] (a) Eder, U.; Sauer, G.; Wiechert. R. *Angew. Chem. Int. Ed.* 1971, 10, 496–497; (b) Hajos, Z. G.; Parrish, D. R.; *J. Org. Chem.* 1974, 39, 1615–1621.

[12] List, B.; Lerner, R. A. Barbas, C. F. *J. Am. Chem. Soc.* 2000, 122, 2395–2396.

[13] Brown, S. P.; Brochu, M.; Sinj, C. J.; MacMillan, D. W. C. *J. Am. Chem. Soc.* 2003, 125, 10808.

[14] Brady, P. B.; Oda, S.; Yamamoto, H. *Organic Lett.* 2014, 16, 3864–3867.

[15] Xu, l.-W.; Luo, J.; Lu, Y. *Chem. Commun.* 2009, 14, 1807–1810

[16] Ooi, T.; Takahashi, M.; Doda, K.; Maruoka, K. *J. Am. Chem. Soc.* 2002, 124, 7640.

[17] Doyle, A. G.; Jacobsen, E. N. *Chem. Rev.* 2007, 113, 5713.

[18] (a) Masson, G.; Housseman, C.; Zhu, J. *Angew. Chem. Int. Ed.* 2007, 119, 4698–4712; (b) Masson, G.; Housseman, C.; Zhu, J. *Angew. Chem. Int. Ed.* 2007, 46, 4614–4628; (c) Aroyan, C. E.; Dermenci, A.; Miller, S. J. *Tetrahedron*, 2009, 65, 4069–4084; (d). Wei. Y.; Shi, M. *Acc. Chem. Res.* 2010, 43, 1005–1018.

[19] (a) Akiyama, T. *Chem. Rev.* 2007, 107, 5744–5758; (b) Rueping, M.; Nachtsheim, B. J.; Ieawsuwan, W.; Atodiresei, I. *Angew. Chem. Int. Ed.* 2011, 123, 6838–6853; (c) Schreiner, P. R. *Chem. Soc. Rev.* 2003, 32, 289–296; (d) Doyle, A. G.; Jacobsen, E. N. *Chem. Rev.* 2007, 107, 5713–5743; (e) Zhang, Z.; Schreiner, P. R. *Chem. Soc. Rev.* 2009, 38, 1187–1198. (f) Aleman, J.; Parra, A.; Jiang, H.; Jorgensen, K. A. *Chem. Eur. J.* 2011, 17, 6890–6899.

[20] (a) Mahlau, M.; List, B. *Angew. Chem.* 2013, 140, 540–556; (b) Mahlau, M.; List, B. *Angew. Chem. Int. Ed.* 2013, 52, 518–533.

[21] Ullrich, S.; Nazir, Z.; Busing, A.; Scheffer, U.; Wirth, D.; Bats, J. W.; Durner, G.; Gobel, M. W. ChemBioChem 2011, 12, 1223.

[22] Chinchilla, R.; Najera, C.; Sanchez-Agullo, P. *Tetrahedron Asymmet*. 1994, 5, 1393–1402.

[23] Iyer, M. S.; Gigstad, K. M.; Namdev, N. D.; Lipton, M. *J. Am. Chem. Soc*. 1996, 118, 4910–4911.

[24] Corey, E. J.; Grogan, M. J. *Org. Lett*. 1999, 1, 157–160.

[25] Kita, T.; Georgieva, A.; Hashimoto, Y.; Nakata, T.; Nagasawa, K. *Angew. Chem. Int. Ed*. 2002, 41, 2832–2834.

[26] Leow, D.; Tan, C.-H. *Synlett* 2010, 11, 1589–1605.

[27] Allingham, M. T.; Howard-Jones, A.; Murphy, P. J.; Thomas, D. A.; Caulkett, P. W. R. *Tetrahedron Lett*. 2003, 44, 8677–8680.

[28] Chen, W.; Yang, W.; Yan, L.; Tan, C.-H.; Jiang, Z. *Chem. Commun*. 2013, 49, 9854–9856.

[29] Mangawa S. K.; Singh, A. K.; Awasthi, S. K. *RSC Adv*. 2015, 5, 61144–61147.

[30] Martinez-Castaneda, A.; Poladura, B.; Rodriguez-Sulla, H.; Concellon, C.; del Amo, V. *Chem. Eur. J*. 2012, 18, 5188–5190.

[31] (a) Genski, T.; Macdonald, G.; Wei, X.; Lewis, N.; Taylor, R. J. K. *ARKIVOC* 2000, (iii), 266–273; (b) McManus, J. C.; Carey, J. S.; Taylor, R. J. K. *Synlett* 2003, 33, 365–368; (c) McManus, J. C.; Genski, T.; Carey, J. S.; Taylor, R. J. K. *Synlett*, 2003, 33, 369–371.

[32] (a) Kita, T.; Shin, B.; Hashimoto, Y.; Nagasawa, K. *Heterocycles*, 2007, 73, 241–247. (b) Shin, B.; Tanaka, S.; Kita, T.; Hashimoto, Y.; Nagasawa, K. *Heterocycles* 2008, 76, 801–810.

[33] (a) Corey, E. J. *Angew. Chem. Int. Ed*. 2002, 114, 1724–1741; (b) Corey, E. J. *Angew. Chem. Int. Ed*. 2002, 114, 1650–1667.

[34] (a) Ma, D.; Cheng, K. *Tetrahedron Asymmet*. 1999, 10, 713–719; (b) Peng, B.; Cheng, K.-J.; Ma, D. *Chin. J. Chem*. 2003, 21, 793–796.

[35] (a) Shen, J.; Nguyen, T. T.; Goh, Y.-P.; Ye, W.; Fu, X.; Tan, C.-H. *J. Am. Chem. Soc*. 2006, 128, 13692–13693; (b) *Eur. J. Org. Chem. 2010, 14, 2635-2655*

[36] Terrasson, V.; de Figueiredo, R. M.; Campagne, J. M. *Eur. J. Org. Chem*. 2010, 14, 2635–2655.

[37] Sohtome,Y.; Shin, B.; Horitsugi, N.; Noguchi, K.; Nagasawa, *Chem. Asian J*. 2011, 6, 2463–2470.

[38] Sasai, H.; Suzuki, T.; Arai, S.; Arai, T.; Shibasaki, M. *J. Am. Chem. Soc.* 1992, 114, 4418–4421.

[39] Takada, K.; Takemura, N.; Cho, K.; Sohtome, Y.; Nagasawa, K. *Tetrahedron Lett.* 2008, 49, 1623–1626.

[40] Fang, B.; Liu, X.; Zhao, J.; Tang, Y.; Lin, L.; Feng, X. *J. Org. Chem.* 2015, 80, 3332–3338.

[41] Ma, D.; Pan, Q.; Han, F. *Tetrahedron Lett.* 2002, 43, 9401–9403.

[42] Howard-Jones, A.; Murphy, P. J.; Thomas, D. A. *J. Org. Chem.* 1999, 64, 1039–1041.

[43] Dai, Q.; Huang, H.; Zhao, J. C.-G. *J. Org. Chem.* 2013, 78, 4153–4157.

[44] Almasi, D.; Alonso, D. A.; Najera, C. *Tetrahedron Asymmet.* 2007, 18, 299–365.

[45] (a) Ishikawa, T.; Araki, Y.; Kumamoto, T.; Seki, H.; Fukuda, K.; Isobe, T. *Chem. Commun.* 2001, 3, 245–246; (b) Ryoda, A.; Yajima, N.; Haga, T.; Kumamoto, T.; Nakanishi, W.; Kawahata, M.; Yamaguchi, K.; Ishikawa, T. *J. Org. Chem.* 2008, 73, 133–141.

[46] Kumamoto, T.; Ebine, K.; Endo, M.; Araki, Y.; Fushimi, Y.; Miyamoto, I.; ishikawa, T.; Isobe, T.; Fukuda, K. *Heterocycles* 2005, 66, 347–359.

[47] Terada, M.; Ube, H.; Yaguchi, Y. *J. Am. Chem. Soc.* 2006, 128, 1454–1455.

[48] (a) Davie, E. A. C.; Mennen, S. M.; Xu. Y.; Miller, S. J. *Chem. Rev.* 2007, 107, 5759–5812; (b) Linton, B. R.; Reutershan, M. H.; Aderman, C. M.; Richardson, E. A.; Brownell, K. R.; Ashley, C. W.; Evans, C. A.; Miller, S. J. *Tetrahedron Lett.* 2007, 48, 1993–1997.

[49] Shen, J.; Tan, C.-H. *Org. Biomol. Chem.* 2008, 6, 4096–4098.

[50] Jiang, Z.; Zhang, Y.; Ye, W.; Tan, C.-H. *Tetrahedron Lett.* 2007, 48, 51–54.

[51] Terada, M.; Ikehara, T.; Ube, H. *J. Am. Chem. Soc.* 2007, 129, 14112–14113.

[52] (a) Saito, N.; Ryodo, A.; Nakanishi, W.; Kumamoto, T.; Ishikawa, T. *Eur. J. Org. Chem.* 2008, 16, 2759–2766; (b) Nising, C. F.; Brase, S. *Chem. Soc. Rev.* 2008, 37, 1218–1228.

[53] (a) Kobayashi, S.; Yazaki, R.; Seki, K.; Yamashita, Y. *Angew. Chem. Int. Ed.* 2008, 120, 1194–1197; (b) Kobayashi, S.; Yazaki, R.; Seki, K.; Yamashita, Y. *Angew. Chem. Int. Ed.* 2008, 47, 5613–5615.

[54] Groger, H. *Chem. Rev.* 2003, 103, 2795–2827.

[55] Castro, A. M. M. *Chem. Rev.* 2004, 104, 2939–3002.

[56] Uyeda, C.; Jacobsen, E. N. *J. Am. Chem. Soc.* 2008, 130, 9228–9229.

[57] Basavaih, D.; Rao, K. V.; Reddy, B. S. *Tetrahedron Asymmet.* 2006, 17, 1036–1040.

[58] Terada, M.; Nakano, M.; Ube, H. *J. Am. Chem. Soc.* 2006, 128, 16044–16045.

[59] (a) Leow, D.; Lin, S.; Chittimalla, S. K.; Fu, X.; Tan, C.-H. *Angew. Chem. Int. Ed.* 2008, 120, 5723–5727; (b) Leow, D.; Lin, S.; Chittimalla, S. K.; Fu, X.; Tan, C.-H. *Angew. Chem. Int. Ed.* 2008, 47, 5641–5645.

[60] Ishikawa, T.; Kumamoto, T. *Synthesis* 2006, 5, 737–752.

[61] Hjelmencrantz, A.; Berg, U. *J. Org. Chem.* 2002, 67, 3585–3594.

[62] Isobe, T.; Fukuda, K.; Araki, Y.; Ishikawa, T. *Chem. Commun.* 2001, 3, 243–244.

Section 2

Carbon Dioxide CO2 Transformation Organocatalyzed

Chapter 5

Organocatalytic Transformation of Carbon Dioxide

Ruimao Hua and Sushmita Roy

Additional information is available at the end of the chapter

http://dx.doi.org/10.5772/63096

Abstract

Catalytic transformation of CO_2 into the value-added organic compounds is a very important and hot research topic in organic synthetic chemistry and green chemistry from the viewpoint of developing CO_2 as C resource. Organocatalytic reactions employing metal-free organic molecules as catalysts have received unprecedented attention in recent years, with the significant advantages of the catalysts being usually inexpensive and stable, and the reactions can be performed under air. This chapter summarizes and gives an overview of the recent advances in the organocatalytic transformation of CO_2 into cyclic carbonates, 2-oxazolidinones, carboxylic derivatives, as well as the synthesis of CO_2-adducts and their application as CO_2 carriers.

Keywords: carbon dioxide, CO_2-adduct, cyclic carbonates, organocatalysis, 2-oxazolidinones

1. Introduction

Carbon dioxide (CO_2) exhibits many good qualities as an ideal C resource in organic synthesis such as non-toxicity, natural abundance, and inexpensiveness. Therefore, a variety of efficient catalyst systems have been developed for the transformation of CO_2 into the useful and value-added organic compounds, even it is a kinetically and thermodynamically stable final product of all combustion processes of organic matters and some comprehensive reviews have been reported [1–3]. On the other hand, the organocatalytic reactions using simple, cheap, stable, and easily available organic compounds as catalysts for various organic transformations have been widely investigated in the past two decades [4–6] and have also played an important tool for catalytic activation of CO_2 and its transformation. In this chapter, we focus on summarizing the representative examples of the recent advancement on the organocatalytic transformation of

CO_2 into the different types of useful molecules, including cyclic carbonates, 2-oxazolidinones, ureas, and carbamates, as well as the CO_2-adducts and their application as CO_2 carriers.

2. Transformation of CO_2 into cyclic carbonates

The coupling of CO_2 with epoxides is an atom-economic transformation for the synthesis of cyclic carbonates, which have high potential application as the aprotic polar solvents [7], electrolytes for lithium ion batteries [8], precursors for organic synthesis [9], and polymers [10].

Ionic liquids (ILs) have been well applied as the efficient organocatalysts in the coupling of CO_2 with terminal epoxides since it was first reported by Deng's group in 2001 using 1-*n*-butyl-3-methylimidazolium (BMIm) and *n*-butylpyridinium (BPy) salts as catalysts [11].

Caló [12] group reported a straightforward method for chemical fixation of CO_2 into terminal epoxides by simply dissolving epoxides in molten tetrabutylammonium bromide and iodide (TBAB and TBAI) as solvent under an atmospheric pressure of CO_2 (**Scheme 1**). The cyclic carbonates could be isolated by vacuum distillation or extraction with organic solvents, and the ionic liquid (IL) was insoluble allowing the recycling of the ammonium salt. In addition, polymerization sensitive epoxides also reacted very well to give the corresponding cyclic carbonates, and the reaction rate depended on the nucleophilicity of the halide ion as well as the structure of the cation. TBAI could be also used as the sole solvent, and at 60°C, the reactions gave the cyclic carbonates in the similar yields.

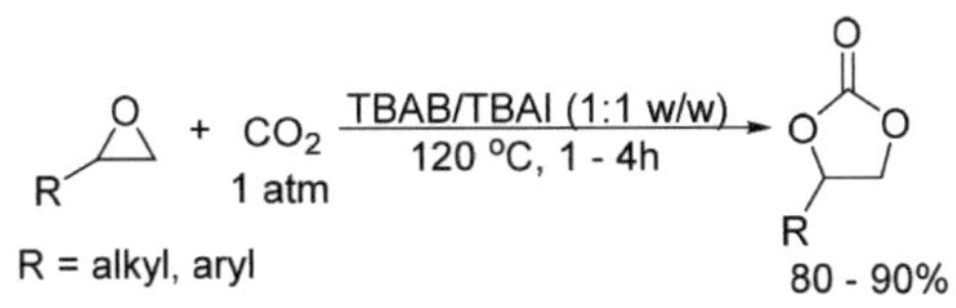

Scheme 1. Formation of cyclic carbonates in tetraalkyammonium salts.

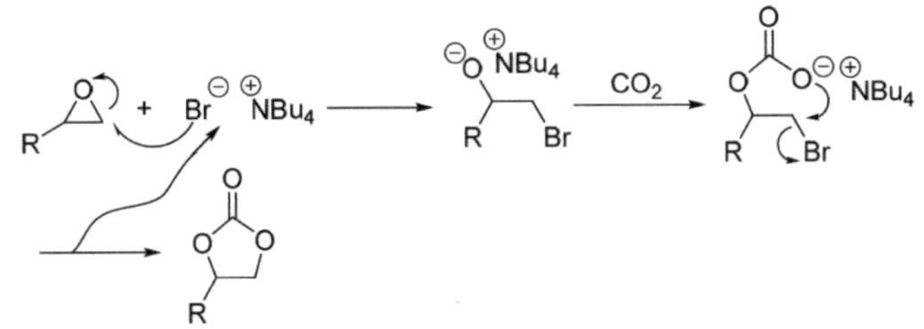

Scheme 2. Plausible mechanism for cyclic carbonate formation in Bu_4NBr.

A plausible mechanism was proposed for the formation of cyclic carbonates including the steps of the ring opening of epoxide by a nuclephilic attack of bromide ion, and the reaction of CO_2 with the oxy anion species (**Scheme 2**).

Another IL such as quaternary ammonium-, phosphonium-, imidazolium-, or pyridinium-based cations with inorganic counter anions have been also used as the efficient catalysts in the synthesis of cyclic carbonate *via* the coupling of CO_2 with epoxides (**Scheme 3**) [13].

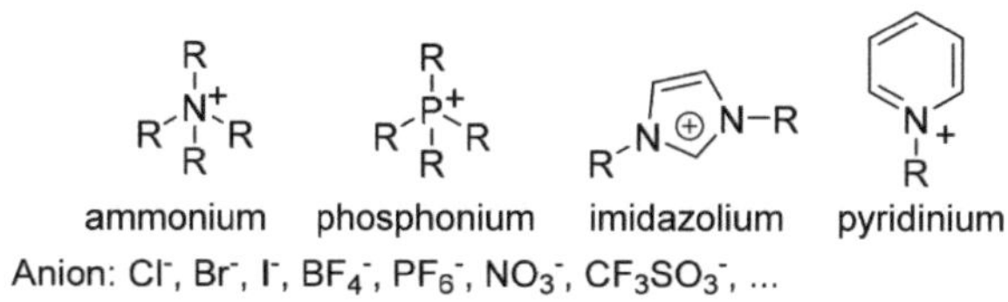

Scheme 3. Ionic liquids as catalysts in the reaction of CO_2 with epoxide.

He's group prepared a series of Lewis basic ILs and examined their catalytic activity in the synthesis of cyclic carbonate from CO_2 and epoxides under solvent-free conditions and established an efficient and recoverable catalyst system using [HDBU]Cl (1,8-diazabicyclo[5.4.0]undec-7-enium chloride) as organocatalyst (**Scheme 4**) [14]. The catalyst system also showed fair catalytic activity to internal cyclohexene oxide.

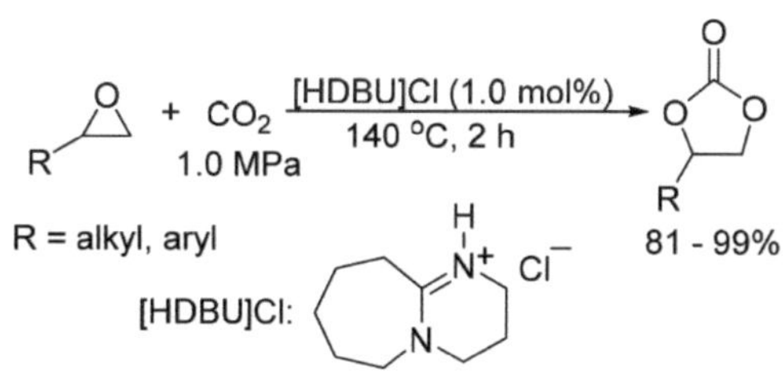

Scheme 4. Formation of cyclic carbonates catalyzed by recoverable [HDBU]CI.

A further work of the same group designed and synthesized a series of polyethylene glycol (PEG)-functionalized basic ILs, and providing the alternative recoverable and high catalytic activity organocatalysts in the coupling of CO_2 with terminal and internal epoxides [15].

In addition, although PPN salts (**Scheme 5, A**) with weak nucleophilic anions such as $PPN^+BF_4^-$ and PPN^+OTf^- were inactive for the coupling of CO_2 with epoxides, PPN^+Cl^- salt was found to be a good organocatalyst for the coupling of CO_2 with neat epoxides without the use of organic solvents to afford cyclic carbonates [16].

Azaphosphatranes as tunable alternative to quaternary ammonium and/or phosphonium catalysts for the synthesis of cyclic carbonates from CO_2 and epoxides was also reported by Martinez and Dufaud's group (**Scheme 5, B**) [17]. In order to examine the nature of the nanospace of the molecular cavity to affect the stability and reactivity of azaphosphatranes as organocatalyst, the same groups further reported the synthesis of supramolecular azaphosphatranes having cavities of different size and shape, and their excellent catalytic activity in the synthesis of cyclic carbonates from CO_2 and epoxides [18].

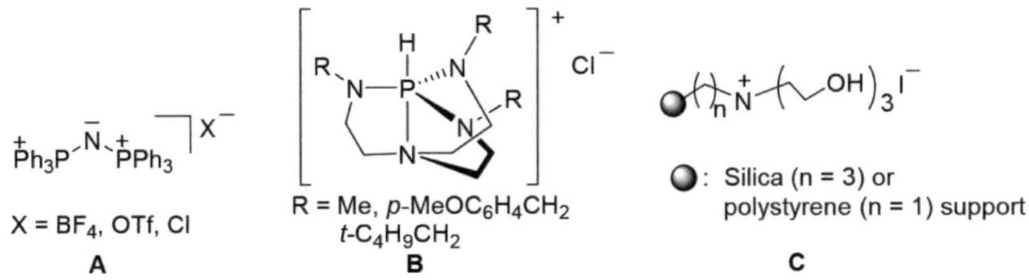

Scheme 5. Organocatalyst for the coupling of CO_2 with epoxides.

Werner's group synthesized a bifunctional ammonium salt covalently bound to a polystyrene or silica support, which showed efficient catalytic activity under solvent-free conditions for the synthesis of cyclic carbonates, developing an alternative recyclable and reusable organocatalyst for the coupling of CO_2 with epoxides (**Scheme 5, C**) [19].

In addition, in order to understand the mechanism of the coupling of CO_2 with epoxides catalyzed by quaternary ammonium salts, Zhang's group studied the mechanism by experimental and density functional theory (DFT). The detailed structural and energetic information about each step of the three elementary steps in the catalytic cycle were obtained, and the effects of the chain length and anion on the reaction mechanisms, as well as the outcomes were also reported [20].

Wong's group designed and synthesized a new IL (**D**), which showed high catalytic activity for the formation of cyclic carbonates under mild conditions in the presence of small amount of water (**Scheme 6**) [21]. The IL plays dual roles as an organocatalyst and the reaction medium. Moreover, IL was very robust under reaction conditions and could be recycled and reused constantly without showing any significant loss in its catalytic activity.

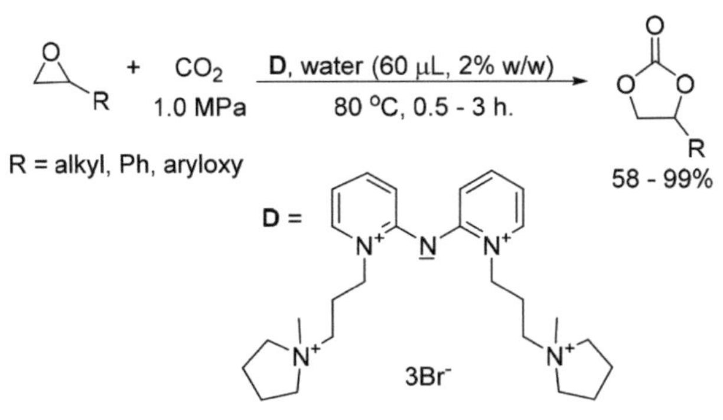

Scheme 6. Ionic liquid as the efficient organocatalyst and reaction medium in cyclic carbonate formation.

In recent years, the simple and cheap organic compounds have been also developed as the efficient organocatalysts in the activation of CO_2 and its transformation to cyclic carbonates.

Shi's group studied the catalytic activity of a combination of phenols with organic bases in the coupling of CO_2 with terminal epoxides and found that *p*-methoxyphenol with 4-dimethylaminopyridine (DMAP) was the best combination to give the cyclic carbonates the excellent yields (**Scheme 7**) [22]. A study of mechanism using *trans*-deuterioethylene oxide as the substrate disclosed that the formation of cyclic carbonate proceeded *via* the epoxy ring activated by phenol by hydrogen bonding and opened by amine (DMAP) and then reacting with CO_2 to give the corresponding cyclic carbonate (**Scheme 8**).

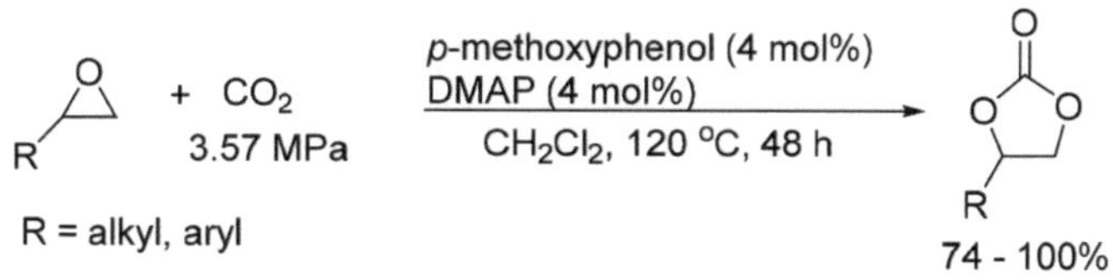

Scheme 7. Reaction of CO_2 with epoxide in the presence of *p*-methoxyphenol and DMAP.

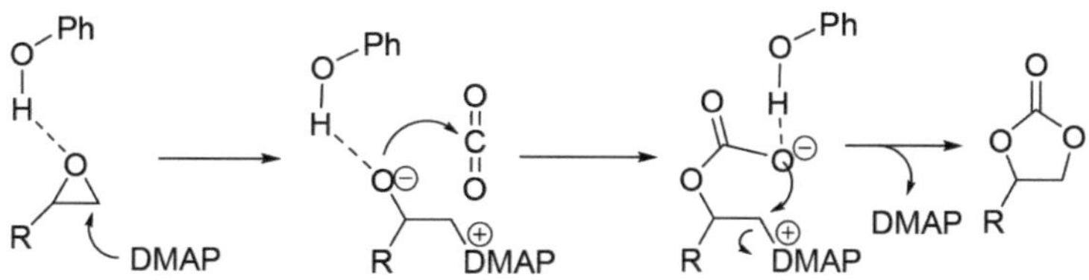

Scheme 8. Proposed mechanism for the formation of cyclic carbonates catalyzed by *p*-methoxyphenol and DMAP.

In addition, Maseras and Kleij's groups examined the catalytic activation of phenols/*n*-Bu_4NI in methyl ethyl ketone as solvent, and optimized pyrogallol/*n*-Bu_4NI was a powerful catalyst with ample substrate scope under very mild reaction conditions (25–45°C, $P(CO_2)$ = 10 bar, 2–5 mol% catalyst) for the preparation of various cyclic carbonates from CO_2 and terminal epoxides [23]. Kleij's group further designed and synthesized the immobilized pyrogallol organocatalyst, developing an efficient and recyclable organocatalyst with a significant advantage of low reaction temperature (45°C) [24]. Furthermore, Kleij's group found that tannic acid, a naturally occurring plant polyphenol, was an efficient organocatalyst with the use of *n*-Bu_4NI as cocatalyst in the same transformation [25].

DMF-scCO_2 system was reported to be a good solvent system for the coupling of 1,2-epoxystyrene with CO_2 affording the corresponding cyclic carbonate [26]. A further investigation by Hua's group disclosed that under solvent-free conditions, the high yields of cyclic carbonates could be achieved by coupling of CO_2 with epoxides in the presence of catalytic amount of DMF [27], and in some cases, the catalytic activity of DMF could be significantly increased by the addition of catalytic amount of H_2O (**Scheme 9**).

Hua's group also investigated the catalytic activity of nitrogen-containing organic compounds, such as amines, anilines, amides, and pyridines in the formation of cyclic carbonates *via* the coupling of CO_2 with terminal epoxides, and confirmed that 2,2′,2″-terpyridine was an excellent organocatalyst to catalyze such type of transformation [28].

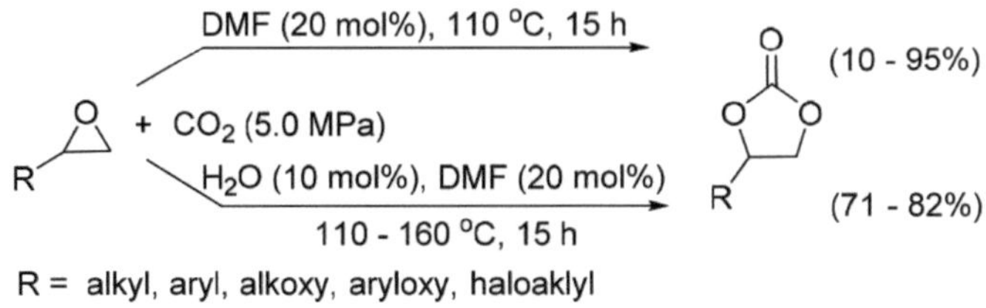

Scheme 9. DMF-catalyzed the formation of cyclic carbonates from CO_2 and epoxides.

On the other hand, the cycloaddition of propargyl alcohols with CO_2 is an efficient and alternative transformation for the formation of α-methylene cyclic carbonates, and Dixneuf's group first reported a PBu_3-catalyzed reaction of tertiary propargyl alcohols with CO_2 in an inert autoclave led to the high yield of the cyclic carbonates (**Scheme 10**) [29]. It was found that in the absence of other solvent, PBu_3 showed higher catalytic activity than PPh_3 and PCy_3.

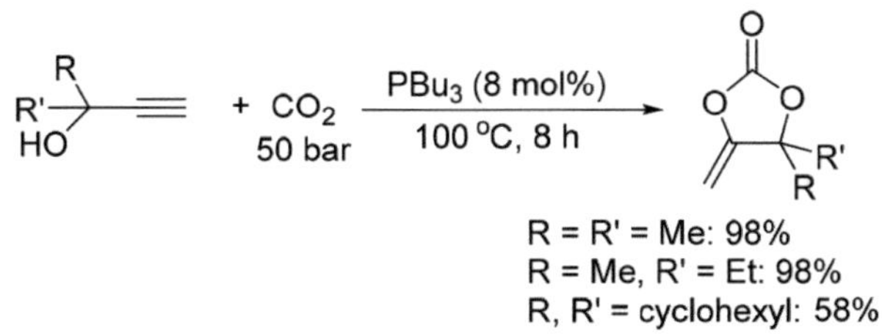

Scheme 10. PBu_3-catalyzed the reaction of tertiary propargyl alcohols with CO_2 affording α-methylene cyclic carbonates.

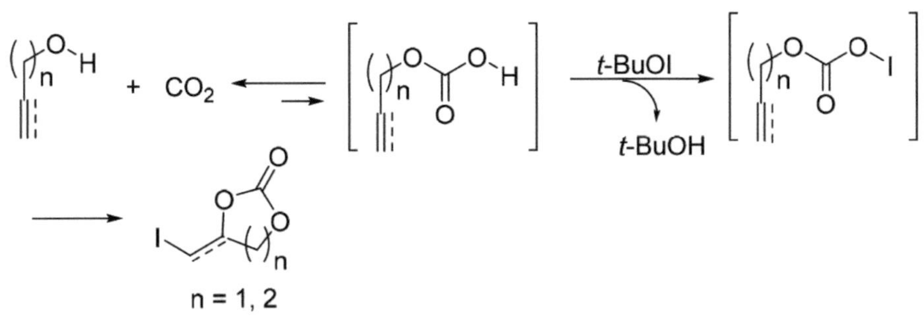

Scheme 11. The formation of iodo-substituted cyclic carbonates.

Minakata's group developed a strategy to offer an innovative approach to the fixation of CO_2 to a wide range of cyclic carbonates *via* the reaction of CO_2 with both olefinic and acetylenic alcohols including the steps of the formation of alkyl carbonic acid and its iodination, as well as the subsequent intramolecular cyclization (**Scheme 11**) [30]. The formation of iodo-substituted cyclic carbonates results in high potential application of the presented strategy in organic synthesis.

http://dx.doi.org/10.5772/63096

Johnston's group also reported a three component reaction of CO_2, homoallylic alcohol and NIS (NIS = *N*-iodosuccinimide, an electrophilic source of iodine) using a chiral BAM catalyst having dual Brønsted acid/base role that presents hydrogen-bond donor and acceptor functionality to activate and adjust substrates in an enantioselective reaction, and cyclic carbonates were obtained enantioselectively (**Scheme 12**) [31].

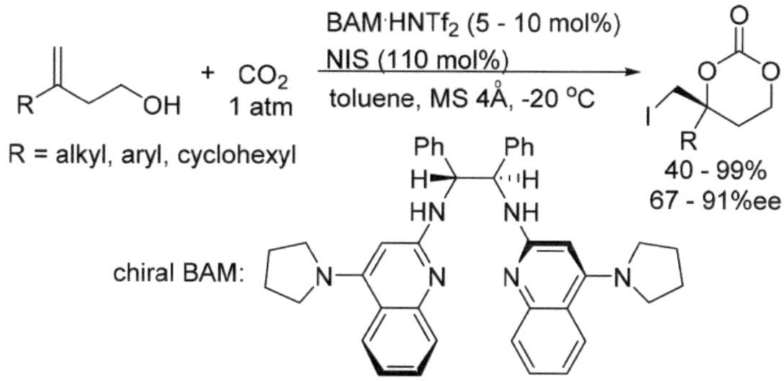

Scheme 12. Cyclic carbonates from homoallylic alcohol, CO_2 and an electrophilic source of iodine.

3. Transformation of CO_2 into 2-oxazolidinones

Substituted 2-oxazolidinones are one of the important five-membered heterocyclic compounds, which not only show interesting biological and physiological activities but also have been applied as starting materials in the synthesis of other functional compounds. The coupling of CO_2 with aziridine, CO_2 with propargylamine, as well as the three-component cycloaddition of CO_2, propargyl alcohol, and primary amine is the most interesting and promising synthetic methods.

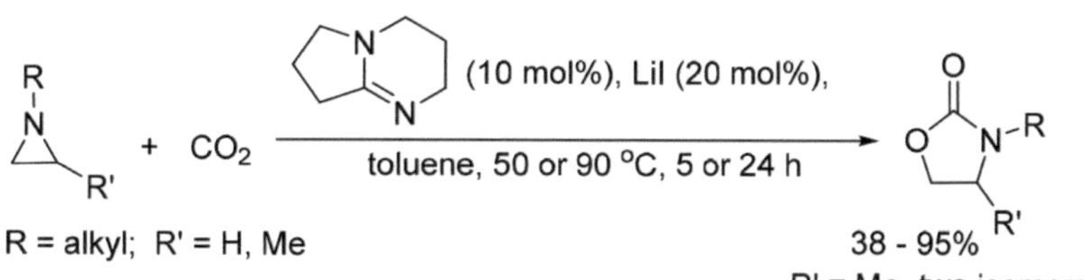

Scheme 13. Synthesis of 2-oxazolidinones from CO_2 and aziridine catalyzed by DBN.

He's group designed and synthesized a series of polyethylene glycol (PEG)-functionalized ionic liquids as recyclable and efficient organocatalysts for selective synthesis of 5-substituted-2-oxazolidinones from the coupling of CO_2 and aziridines. It was found that $PEG_{6000}(NBu_3Br)_2$ (PEG MW6000) [32] and BrDBN-PEG_{150}-DBNBr (DBN: 1,5-diazabicyclo[4.3.0]non-5-ene; PEG MW150) [33] were the efficient catalyst not only affording the expected 5-substituted-2-oxazolidinones in good yields, but also showing excellent regioselectivities. The same group also developed a proline-catalyzed synthesis of 5-aryl-2-oxazolidinones from CO_2 and aziridines under solvent-free conditions [34].

Liu's group developed an efficient catalytic system using DBN as organocatalyst and LiI as an additive under atmospheric pressure of CO_2 in toluene to catalyze the coupling of CO_2 with aziridines giving 2-oxazolidinones (**Scheme 13**) [35]. The procedure was tolerated by a number of *N*-alkyl aziridines bearing various functional groups in alkyl terminal position, but *N*-tosyl aziridines did not undergo the coupling reaction probably due to the less nucleophilic activity of the nitrogen. In addition, the formation of the DBN-CO_2 was proposed to be the intermediate of the catalytic cycle (**Scheme 14**).

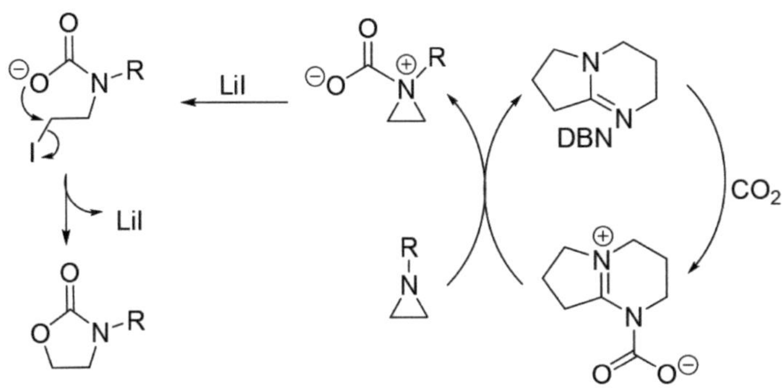

Scheme 14. Proposed mechanism in the synthesis of 2-oxazolidinones from CO_2 and aziridines catalyzed by DBN.

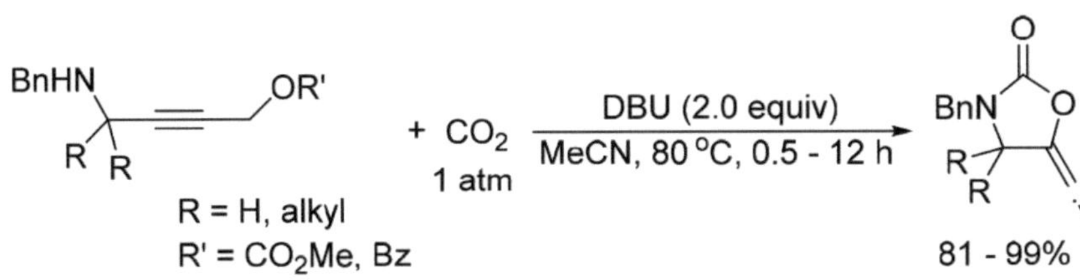

Scheme 15. DBU-mediated synthesis of substituted 5-vinylideneoxazolidin-2-ones.

Yoshida and Ihara's groups investigated the reaction of 4-(benzylamino)-2-butynyl carbonates and benzoates with an atmospheric pressure of CO_2 in the presence of DBU (1,8-diazabicyclo[5.4.0]undec-7-ene), developing a DBU-mediated synthesis of substituted 5-vinylideneoxazolidin-2-ones, which are attractive and important compounds in both medicinal and synthetic organic chemistry (**Scheme 15**) [36].

The three-component cycloaddition of CO_2, propargyl alcohol and primary amine catalyzed by organocatalyst affording 4-methylene-2-oxazolidinones was first reported by Dixbeuf's group with the use of simple and cheap PBu_3 as catalyst and use of an excess amount of tertiary propargylic alcohols (**Scheme 16**) [37].

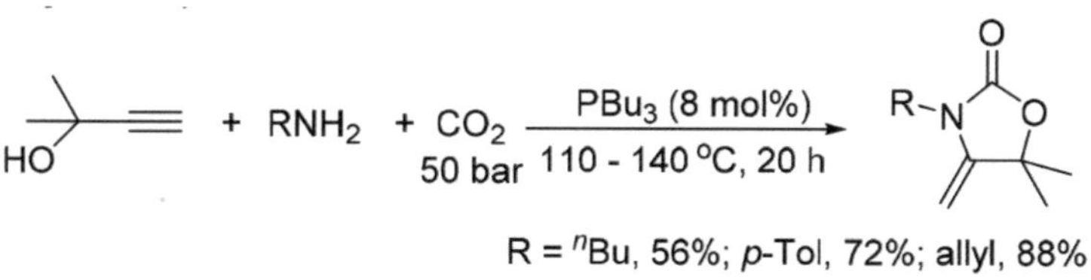

Scheme 16. PBu_3-catalyzed three-component cycloaddition of CO_2, propargyl alcohol and primary amine affording 4-methylene-2-oxazolidinones.

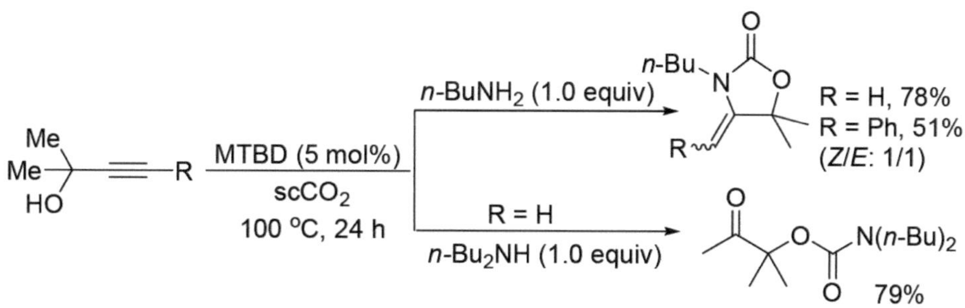

Scheme 17. Reaction of propargyl alcohol with $scCO_2$ in the presence of MTBD.

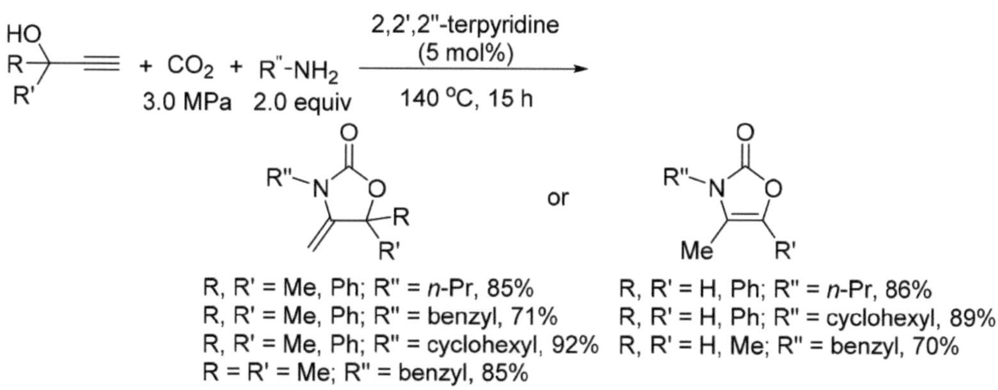

Scheme 18. Synthesis of 4-methylene-2-oxazolidinones or 2(3*H*)-oxazolones catalyzed by 2.2′.2″-terpyridine.

Costa's group studied the synthesis of cyclic carbonates or carbamates and oxalkyl carbonates or carbamates *via* the direct incorporation of CO_2 into propargyl alcohol using either $scCO_2$ as solvent and reagent or gaseous CO_2 in MeCN as the solvent in the presence of a variety of organic bases. It was found that bicyclic guanidines, such as MTBD (1,3,4,6,7,8-hexaydro-1-methyl-2*H*-pyrimido[1,2-*a*]pyrimidine), TBD (2,3,4,6,7,8-hexaydro-1*H*-pyrimido [1,2-*a*]pyrimidine), and TBD-pol (1,3,4,6,7,8-hexaydro-2*H*-pyrimido [1,2-*a*]pyrimidine supported on polystyrene), could effectively catalyzed CO_2 transformation to the different carbonyl compounds depending upon the use of external nucleophiles [38]. As shown in **Scheme 17**, in

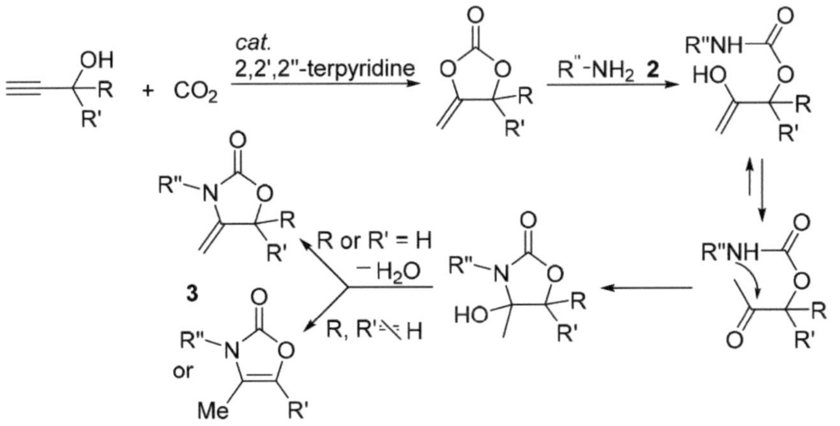

Scheme 19. Proposed mechanism for the formation of 4-methylene-2-oxazolidinones and 2(3*H*)-oxazolones.

the presence of MTBD, the reactions of propargyl alcohols with primary and secondary amines selectively afforded α-methyleneoxazolidinones and acyclic carbamates in good yields, respectively.

Hua's group also investigated the three-component cycloaddition of CO_2, propargyl alcohol and primary amine in the presence of pyridine and its derivatives as organocatalysts under solvent-free conditions, and 2,2′,2″-Terpyridine was found to be the efficient organocatalyst to afford 4-methylene-2-oxazolidinones or 2(3*H*)-oxazolones in good to high yields depending on the structures of propargyl alcohols (**Scheme 18**). A proposed mechanism for the formation of 4-methylene-2-oxazolidinones and 2(3*H*)-oxazolones is depicted in **Scheme 19**, it includes the formation of α-methylene cyclic carbonate as intermediate, and its nucleophilic addition reaction with primary amine to give N-alkylcarbamate, subsequent cyclization by intramolecular nucleophilic addition reaction and dehydration reaction affording 4-methylene-2-oxazolidinones or 2(3*H*)-oxazolones, depending upon the substituents R and R′.

In addition, 2,2′,2″-terpyridine also showed high catalytic activity in the coupling of CO_2 with aziridines bearing either electron-donating or electron-withdrawing *N*-substitutents to give substituted 2-oxazolidinones in high yields. Therefore, 2,2′,2″-terpyridine was applied as an efficient organocatalyst in the transformation of CO_2 into not only cyclic carbonates [28], but also 2-oxazolidinones and 2(3*H*)-oxazolones [39].

In addition, ILs were also reported to be the effective promoter and reaction media for the synthesis of 4-methylene-2-oxazolidinones from CO_2, propargyl alcohol, and amines with high yields [40].

4. Transformation of CO_2 into carboxylic derivatives

CO_2 is one of the good candidates in the synthesis of carboxylic derivatives. Nitrogen-containing organic bases mediated the formation of diarylureas *via* the reaction of CO_2 with aromatic amines was a well-known procedure [41–42]. It has been also known that carbamic acids derived from the reaction of amines with CO_2 gas can be transferred into isocyanates, and then ureas and carbamates *via* the further reactions with amines or alcohols. Peterson's group reported the parallel synthesis of ureas and carbamates from CO_2 and amines catalyzed by DBU and in the presence of PBu_3/DBAD (Mitsunobu reagent [43]; DBAD: di-^{t}butylazodicarboxylate) (**Scheme 20**) [44]. It was proposed that carbamic acids derived from primary amines reacted with Mitsunobu reagent to generate isocyanates in situ, which were condensed with primary and secondary amines to afford the expected unsymmetrical di- and trisubstituted ureas. Similarly, carbamic acids from secondary amines reacted with alcohols activated with Mitsunobu reagents to form carbamates *via* an S_N2 mechanism.

Skrydstrup's group investigated the CO_2 trapping with 2-alkynyl indoles in the presence of various organic bases and developed an efficient TBD-catalyzed the cycloaddition of CO_2 with a variety of substituted 2-alkynyl indoles to afford tricyclic indole-containing ring compounds, good results were obtained with aromatic, heteroaromatic, and aliphatic 2-alkynyl indoles in terms of both yields and selectivities (**Scheme 21**) [45]. The new methodology developed a procedure for the formation of C-C bond between CO_2 and an indole derivative catalyzed by an organocatalyst.

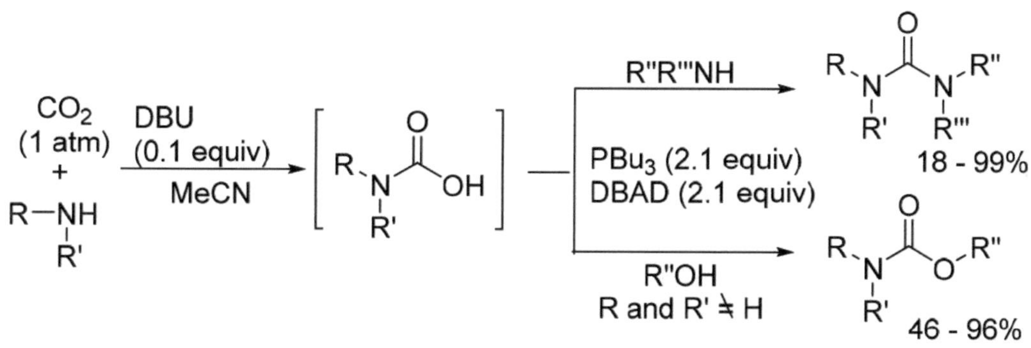

Scheme 20. Parallel synthesis of urea and carbamate from CO_2 and amine via carbamic acid.

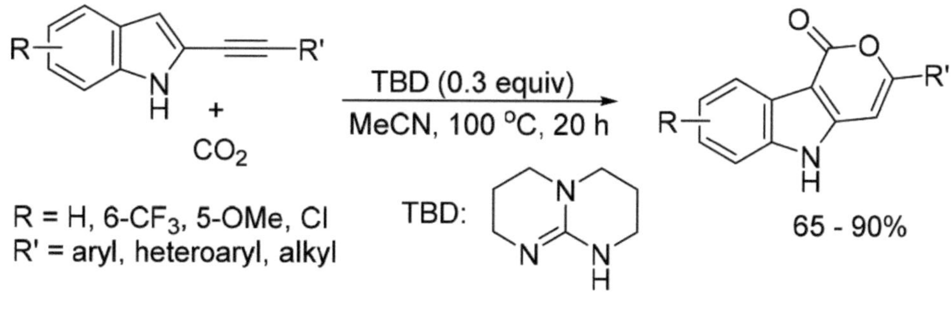

Scheme 21. TBD-catalyzed cycloaddition of CO_2 with 2-alkynyl indoles.

5. CO_2-adduct and its use as precatalyst in CO_2 transformation

CO_2 is a typical electrophilic reagent, and the synthesis and application of its adduct with nucleophiles have been considered to be an efficient way for CO_2 capture, activation, and further transformation.

DBU-CO_2 adduct could be prepared and isolated as a white powder in good yield by the reaction of DBU with CO_2 in anhydrous acetonitrile at 5°C and was first used as the efficient carrier of CO_2 in the synthesis of *N*-alkyl carbamates by a transcarboxylation of amines and subsequent *O*-alkylation using ethyl iodide [46].

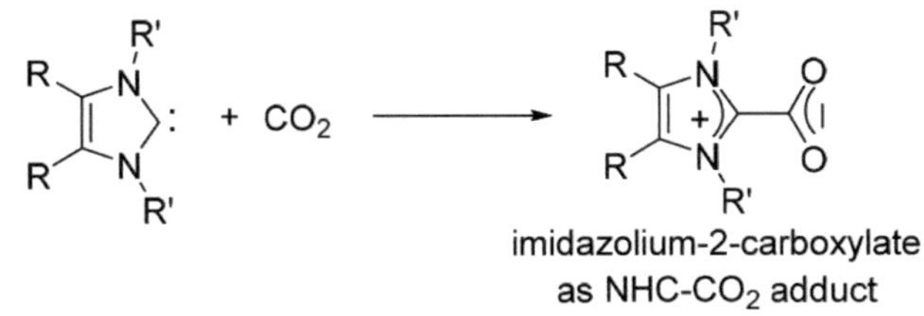

Scheme 22. Formation of NHC–CO_2 adduct.

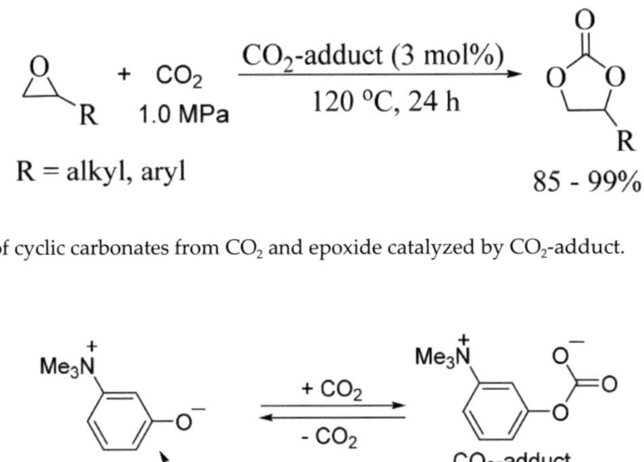

Scheme 23. Formation of cyclic carbonates from CO_2 and epoxide catalyzed by CO_2-adduct.

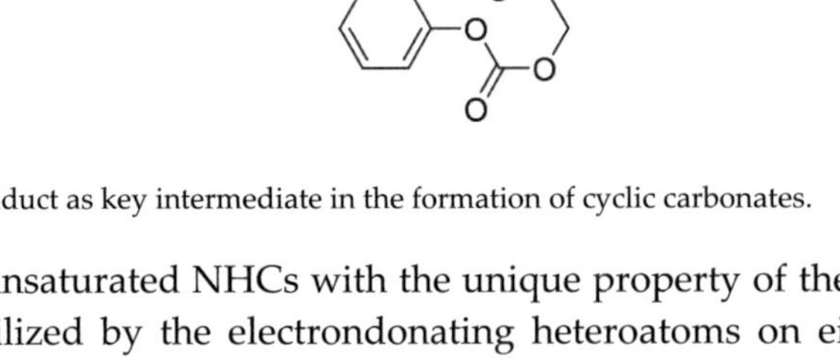

Scheme 24. Betaine-CO_2 adduct as key intermediate in the formation of cyclic carbonates.

On the other hand, unsaturated NHCs with the unique property of the carbon atom having strong basicity, stabilized by the electrondonating heteroatoms on either side, have been applied as versatile ligands in transition metal complexes and organocatalysts [47]. NHCs have been found to react easily with CO_2 by its nucleophilic addition to C=O bond as the key step, resulting in the formation of carboxylates, a NHC-CO_2 adduct proposed as carriers of NHC as well as CO_2 (**Scheme 22**) [48–50].

Sakai's group designed and synthesized bifunctional organocatalysts bearing an ammonium betaine framework, which showed high catalytic activation of CO_2, and catalyze the coupling of CO_2 with terminal epoxides affording cyclic carbonates in good yields (**Scheme 23**) [51]. Among them, 3-(trimethylammonio)phenolate was found to be one of the most active organocatalysts, and the formation of betaine-CO_2 adduct was demonstrated to be the key intermediate (**Scheme 24**).

Moreover, several thermally stable CO_2 adducts of *N*-heterocyclic carbenes (NHC–CO_2) from the reaction of CO_2 with NHCs were found to be the efficient organocatalysts in the coupling of CO_2 with epoxides [52].

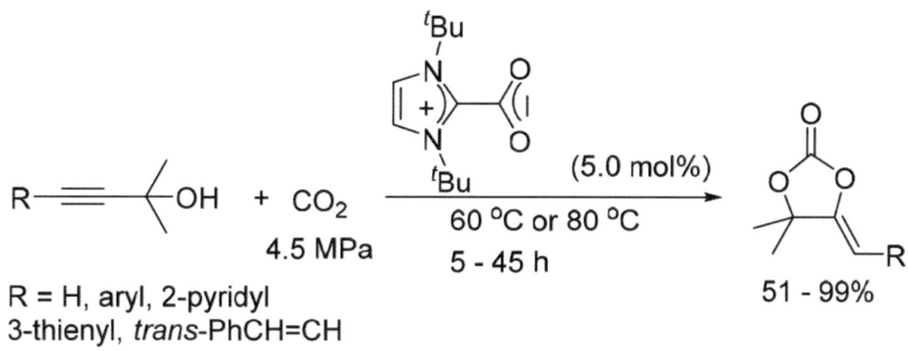

Scheme 25. Formation of 4-methylene cyclic carbonates catalyzed by NHC–CO_2 adduct.

In addition, Ikariya's group synthesized several 1,3-dialkylimidazolium-2-carboxylates (NHC-CO_2 adduct) and investigated their catalytic activity in the cycloaddition reaction of CO_2 with propargyl alcohols affording 4-methylene cyclic carbonates (**Scheme 25**) [53]. 1,3-di-*t*-butylimidazolium-2-carboxylate showed high catalytic activity under solvent-free conditions for the formation of the desired cyclic carbonates, and the reaction catalyst tolerated substrates bearing heterocycles such as pyridine and thiophene. The substrate having an olefinic group at the acetylenic terminus also provided the desired 5-exo-dig cyclization product in good yield. In addition, the C=C double bond at 4-position was found to have a *Z* configuration as determined by NMR spectroscopy, indicating that the addition to the alkynes proceeded predominantly in a *trans* fashion. Moreover, the catalyst was also found to be successful in the cyclic carbonate synthesis *via* the coupling of CO_2 with epoxides.

1-*n*-butyl-3-methylimidazolium-2-carboxylate was also applied as organocatalyst in the synthesis of glycerol carbonate *via* transesterification of glycerol with dimethyl carbonate (DMC) [54].

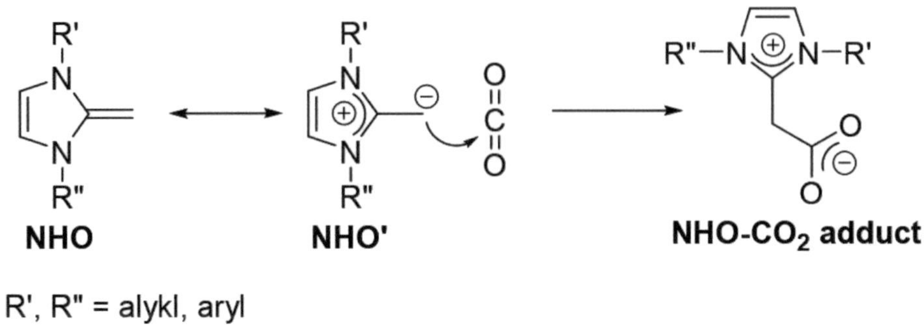

Scheme 26. Formation of NHC–CO_2 adduct.

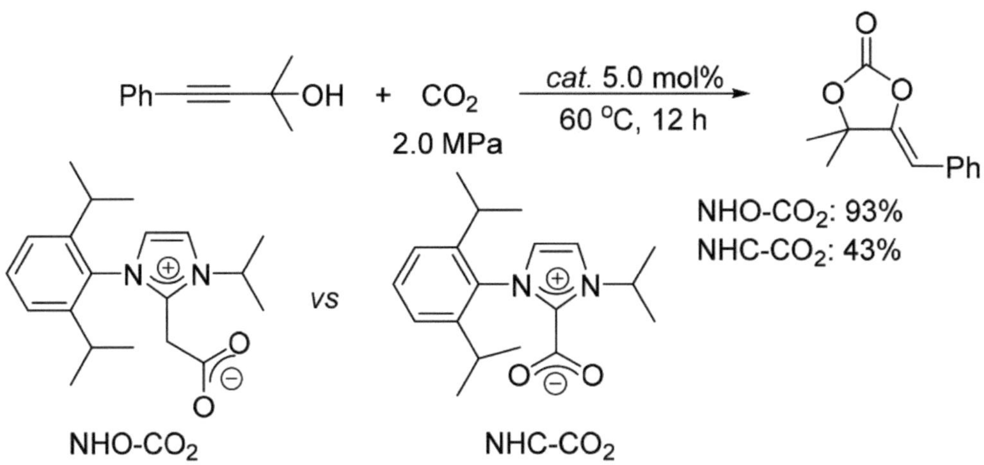

Scheme 27. Formation of 4-methylene cyclic carbonates catalyzed by NHC–CO_2 and NHC–CO_2 adducts.

N-heterocyclic olefin (NHO), shown in **Scheme 26**, is considered to be advantageous to stabilize a positive charge due to the aromatization of the heterocyclic ring resulting in the terminal carbon atom of the olefins more electronegative (NHO′). Therefore, NHO was considered as potent nucleophile for CO_2 capture, activation, and further transformation. Lu's group first synthesized a variety of NHO-CO_2 adducts and studied their catalytic activity in the reaction of CO_2 with propargyl alcohols giving 4-methylene cyclic carbonates (**Scheme 27**) [55]. It was found that NHO-CO_2 not only showed high catalytic activity for the transformation, but also showed higher catalytic activity compared to the corresponding NHC-CO_2 adducts at the same reaction conditions. The higher activity of NHO-CO_2 adduct was tentatively attributed to its low stability for easily releasing the CO_2 moiety and/or the desired product in a possible rate limiting step in the catalytic cycle.

The same group also reported the synthesis of various CO_2, COS, and CS_2 adducts of NHO, and these adducts were found to be efficient in catalyzing the cycloaddition reaction of CO_2

with epoxides to selectively afford the corresponding cyclic carbonates. Among them, NHO-CO_2 adducts were found to be more active [56]. Furthermore, a variety of CO_2 adducts of phosphorus ylides were prepared by the same group, and they were demonstrated to be the highly active organocatalysts for CO_2 transformation under mild conditions to cyclic carbonates, oxazolidinone, N-methylated, and N-formylated amines [57].

6. Conclusion

The use of CO_2 as a C starting material for the synthesis of useful and value-added organic compounds is an important and challenge research topic in the academic and industrial interest. The representative examples summarized in this chapter suggest that the simple, easily available, oxygen- and moisture-tolerated organocatalysts have played an important role in developing the promising and practical catalysis for the transformation of CO_2 to various organic compounds. It seems reasonable to expect that the organocatalyzed CO_2 transformation to much more different types of functional organic compounds will be greatly developed with the inspiration of the reported innovative progress. For example, Cantat's group recently developed the novel and interesting organocatalyst systems for the transformation of CO_2 to methylene in the synthesis of aminal derivatives, and CO_2 as CO source in the formylation of amines using hydrosilanes as reductants catalyzed by nitrogen-containing organic bases [58–59]. The simple accessibility of CO_2 and the vast range of possibilities to introduce various functional groups are some of the attractive features of CO_2 transformations.

Author details

Ruimao Hua* and Sushmita Roy

*Address all correspondence to: ruimao@mail.tsinghua.edu.cn

Department of Chemistry, Tsinghua University, Beijing, China

References

[1] Aresta M, Dibeneddetto A. Utilization of CO_2 as a chemical feedstock: opportunities and challenges. Dalton Trans. 2007;2975–2992. DOI: 10.1039/b700658f

[2] Sakakura T, Choi JC, Yasuda H. Transformation of carbon dioxide. Chem. Rev. 2007;107:2365–2387. DOI: 10.1021/cr068357u

[3] Wang JL, Miao CX, Dou XY, Gao J, He LN. Carbon dioxide in heterocyclic synthesis. Curr. Org. Chem. 2011;15:621–646. DOI: 10.2174/138527211794518952

[4] Bertelsen S, Jørgensen KA. Organocatalysis: after the gold rush. Chem. Soc. Rev. 2009;38:2178–2189. DOI: 10.1039/b903816g

[5] Alemán J, Cabrera S. Applications of asymmetric organocatalysis in medicinal chemistry. Chem. Soc. Rev. 2013;42:774–793. DOI: 10.1039/c2cs35380f

[6] Kiesewetter MK, Shin EJ, Hedrick JL, Waymouth RM. Organocatalysis: opportunities and challenges for polymer synthesis. Macromolecules 2010;43:2093–2107. DOI: 10.1021/ma9025948

[7] Bayardon J, Holz J, Schäffner B, Andrushko V, Verevkin S, Preetz A, Börner A. Propylene carbonate as a solvent for asymmetric hydrogenations. Angew. Chem. Int. Ed. 2007;46:5971–5974. DOI: 10.1002/anie.200700990

[8] Tsuda T, Kondo K, Tomioka T, Takahashi Y, Matsumoto H, Kuwabata S, Hussey CL. Design, synthesis, and electrochemistry of room-temperature ionic liquids functionalized with propylene carbonate. Angew. Chem. Int. Ed. 2011;50:1310–1313. DOI: 10.1002/anie.201005208

[9] Shaikh AAG, Sivaram S. Organic carbonates. Chem. Rev. 1996;96:951–976. DOI: 10.1021/cr950067i

[10] Lu XB, Darensbourg DJ. Cobalt catalysts for the coupling of CO_2 and epoxides to provide polycarbonates and cyclic carbonates. Chem. Soc. Rev. 2012;41:1462–1484. DOI: 10.1039/c1cs15142h

[11] Peng J, Deng Y. Cycloaddition of carbon dioxide to propylene oxide catalyzed by ionic liquids. New J. Chem. 2001;25:639–641. DOI: 10.1039/b008923k

[12] Caló V, Nacci A, Monopoli A, Fanizzi A. Cyclic carbonate formation from carbon dioxide and oxiranes in tetrabutylammonium halides as solvents and catalysts. Org. Lett. 2002;4:2561–2563. DOI: 10.1021/ol026189w

[13] Sun J, Fujita S, Arai M. Development in the green synthesis of cyclic carbonate from carbon dioxide using ionic liquids. J. Organomet. Chem. 2005;690:3490–3497. DOI: 10.1016/j.jorganchem.2005.02.011

[14] Yang ZZ, He LN, Miao CX, Chanfreau S. Lewis basic ionic liquids-catalyzed conversion of carbon dioxide to cyclic carbonates. Adv. Synth. Catal. 2010;352:2233–2240. DOI: 10.1002/adsc.201000239

[15] Yang ZZ, Zhao YN, He LN, Gao J, Yin ZS. Highly efficient conversion of carbon dioxide catalyzed by polyethylene glycol-functionalized basic ionic liquids. Green Chem. 2012;14:519–527. DOI: 10.1039/c2gc16039k

[16] Sit WN, Ng SM, Kwong KY, Lau CP. Coupling reactions of CO_2 with neat epoxides catalyzed by PPN salts to yield cyclic carbonates. J. Org. Chem. 2005;70:8583–8586. DOI: 10.1021/jo051077e

[17] Chatelet B, Joucla L, Dutasta JP, Martinez A, Szeto KC, Dufaud, V. Azaphosphatranes as structurally tunable organocatalysts for carbonate synthesis from CO_2 and epoxides. J. Am. Chem. Soc. 2013;135:5348–5351. DOI: 10.1021/ja402053d

[18] Chatelet B, Joucla L, Dutasta JP, Martinez A, Dufaud V. Azaphosphatrane organocatalysts in confined space: cage effect in CO_2 conversion. Chem. Eur. J. 2014;20:8571–8574. DOI: 10.1002/chem.201402058

[19] Kohrt C, Werner T. Recyclable bifunctional polystyrene and silica gel-supported organocatalyst for the coupling of CO_2 with epoxides. ChemSusChem 2015;8:2031–2034. DOI: 10.1002/cssc.201500128

[20] Wang JQ, Dong K, Cheng WG, Sun J, Zhang SJ. Insights into quaternary ammonium salts-catalyzed fixation carbon dioxide with epoxides. Catal. Sci. Technol. 2012;2:1480–1484. DOI: 10.1039/C2CY20103H

[21] Wong ML, Chan PH, Zhou ZY, Lee KH, Cheung KC, Wong KY. A robust ionic liquid as reaction medium and efficient organocatalyst for carbon dioxide fixation. ChemSusChem 2008;1:67–70. DOI: 10.1002/cssc.200700097

[22] Shen YM, Duan WL, Shi M. Phenol and organic bases co-catalyzed chemical fixation of carbon dioxide with terminal epoxides to form cyclic carbonates. Adv. Synth. Catal. 2003;345:337–340. DOI: 10.1002/adsc.200390035

[23] Whiteoak CJ, Nova A, Maseras F, Kleij AW. Merging sustainability with organocatalysis in the formation of organic carbonates by using CO_2 as a feedstock. ChemSusChem 2012;5:2032–2038. DOI: 10.1002/cssc.201200255

[24] Whiteoak CJ, Henseler AH, Ayats C, Kleij, AW, Pericàs, MA. Conversion of oxiranes and CO_2 to organic cyclic carbonates using a recyclable, bifunctional polystyrene-supported organocatalyst. Green Chem. 2014;16:1552–1559. DOI: 10.1039/c3gc41919c

[25] Sopeña S, Fiorani G, Martín C, Kleij AW. Highly efficient organocatalyzed conversion of oxiranes and CO_2 into organic carbonates. ChemSusChem 2015;8:3248–3254. DOI: 10.1002/cssc.201500710

[26] Kawanami H, Ikushima Y. Chemical fixation of carbon dioxide to styrene carbonate under supercritical conditions with DMF in the absence of any additional catalysts. Chem. Commun. 2000;2089–2090. DOI: 10.1039/b006682f

[27] Jiang JL, Hua R. Efficient DMF-catalyzed coupling of epoxides with CO_2 under solvent-free conditions to afford cyclic carbonates. Synth. Commun. 2006;36:3141–3148. DOI: 10.1080/00397910600908744

[28] Liu H, Zeng R, Hua R. 2,2′,2″-Terpyridine-catalyzed synthesis of cyclic carbonates from epoxides and carbon dioxide under solvent-free conditions. Int. J. Mol. Sci. 2014;15:9945–9951. DOI: 10.3390/ijms15069945

[29] Fournier J, Bruneau C, Dixneuf PH. Phosphine catalysed synthesis of unsaturated cyclic carbonates from carbon dioxide and propargylic alcohols. Tetrahedron Lett. 1989;30:3981–3982. DOI: 10.1016/S0040-4039(00)99300-6

[30] Minakata S, Sasaki I, Ide T. Atmospheric CO_2 fixation by unsaturated alcohols using *t*BuOI under neutral conditions. Angew. Chem. Int. Ed. 2010;49:1309–1311. DOI: 10.1002/anie.200906352

[31] Vara BA, Struble TJ, Wang W, Dobish MC, Johnston JN. Enantioselective small molecule synthesis by carbon dioxide fixation using a dual brønsted acid/base organocatalyst. J. Am. Chem. Soc. 2015;137:7302–7305. DOI: 10.1021/jacs.5b04425

[32] Du Y, Wu Y, Kiu AH, He LN. Quaternary ammonium bromide functionalized polyethylene glycol: a highly efficient and recyclable catalyst for selective synthesis of 5-aryl-2-oxazolidinones from carbon dioxide and aziridines under solvent-free conditions. J. Org. Chem. 2008;73:4709–4712. DOI: 10.1021/jo800269v

[33] Zhao YN, Yang ZZ, Luo SH, He LN. Design of task-specific ionic liquids for catalytic conversion of CO_2 with aziridines under mild conditions. Catal. Today 2013;200:2–8. DOI: 10.1016/j.cattod.2012.04.006

[34] Dou XY, He LN, Yang ZZ. Proline-catalyzed synthesis of 5-aryl-2-oxazolidinones from carbon dioxide and aziridines under solvent-free conditions. Synth. Commun. 2012;42:62–74. DOI: 10.1080/00397911.2010.521903

[35] Wu Y, Liu G. Organocatalyzed cycloaddition of carbon dioxide to aziridines. Tetrahedron Lett. 2011;52:6450–6452. DOI: 10.1016/j.tetlet.2011.09.092.

[36] Yoshida M, Komatsuzaki Y, Ihara M. Synthesis of 5-vinylideneoxazolidin-2-ones by DBU-mediated CO_2-fixation reaction of 4-(benzylamino)-2-butynyl carbonates and benzoates. Org. Lett. 2008;10:2083–2086. DOI: 10.1021/ol800663v

[37] Fournier J, Bruneau C, Dixneuf PH. A simple synthesis of oxazolidinones in one step from carbon dioxide. Tetrahedron Lett. 1990;31:1721–1722. DOI: 10.1016/S0040-4039(00)88863-2

[38] Ca' ND, Gabriele B, Ruffolo G, Veltri L, Zanetta T, Costa M. Effective guanidine-catalyzed synthesis of carbonate and carbamate derivatives from propargyl alcohols in supercritical carbon dioxide. Adv. Synth. Catal. 2011;353:133–146. DOI: 10.1002/adsc.201000607

[39] Liu H, Hua R. Conversion of carbon dioxide into 2-oxazolidinones and 2(3*H*)-oxazolones catalyzed by 2,2′,2″-terpyridine. Tetrahedron 2016;72:1200–1204. DOI: 10.1016/j.tet.2016.01.015

[40] Zhang Q, Shi F, Gu Y, Yang J, Deng Y. Efficient and eco-friendly process for the synthesis of *N*-substituted 4-methylene-2-oxazolidinones in ionic liquids. Tetrahedron Lett. 2005;46:5907–5911. DOI: 10.1016/j.tetlet.2005.06.116

[41] Cooper CF, Falcone SJ. A simple one-pot procedure for preparing symmetrical diarylureas from carbon dioxide and aromatic amines. Synth. Commun. 1995;25:2467–2474. DOI: 10.1080/00397919508015452

[42] Yamazaki N, Higashi F, Iguchi T. Carbonylation of amines with carbon dioxide under atmospheric conditions. Tetrahedron Lett. 1974;15:1191–1194. DOI: 10.1016/S0040-4039(01)82442-4

[43] Horvath MJ, Saylik D, Elmes PS, Jackson WR, Lovel CG, Moody K. A Mitsunobu-based procedure for the preparation of alkyl and hindered aryl isocyanates from primary amines and carbon dioxide under mild conditions. Tetrahedron Lett. 1999;40:363–366. DOI: 10.1016/S0040-4039(98)02312-0

[44] Peterson SL, Stucka SM, Dinsmore CJ. Parallel synthesis of ureas and carbamates from amines and CO_2 under mild conditions. Org. Lett. 2010;12:1340–1343. DOI: 10.1021/ol100259j

[45] Xin Z, Lescot C, Friis SD, Daasbjerg K, Skrydstrup T. Organocatalyzed CO_2 trapping using alkynyl indoles. Angew. Chem. Int. Ed. 2015;54:6862–6866. DOI: 10.1002/anie.201500233

[46] Pérez ER, da Silva MO, Costa VC, Rodrigues-Filho UP, Franco DW. Efficient and clean synthesis of N-alkyl carbamates by transcarboxylation and O-alkylation coupled reactions using a DBU–CO_2 zwitterionic carbamic complex in aprotic polar media. Tetrahedron Lett. 2002;43:4091–4093. DOI: 10.1016/S0040-4039(02)00697-4

[47] Hopkinson MN, Richter C, Schedler M, Glorius S. An overview of *N*-heterocyclic carbenes. Nature. 2014;510:485–496. DOI: 10.1038/nature13384

[48] Holbrey JD, Reichert WM, Tkatchenko I, Bouajila E, Walter O, Tommasi I, Rogers RD. 1,3-Dimethylimidazolium-2-carboxylate: the unexpected synthesis of an ionic liquid precursor and carbene-CO_2 adduct. Chem. Commun. 2003;28–29. DOI: 10.1039/b211519k

[49] Duong HA, Tekavec TN, Arif AM, Louie J. Reversible carboxylation of N-heterocyclic carbenes. Chem. Commun. 2004;112–113. DOI: 10.1039/b311350g

[50] Voutchkova AM, Feliz M, Clot E, Eisenstein O, Crabtree RH. Imidazolium carboxylates as versatile and selective N-heterocyclic carbene transfer agents: synthesis, mechanism, and applications. J. Am. Chem. Soc. 2007;129:12834–12846. DOI: 10.1021/ja0742885

[51] Tsutsumi Y, Yamakawa K, Yoshida M, Ema T, Sakai T. Bifunctional organocatalyst for activation of carbon dioxide and epoxide to produce cyclic carbonate: betaine as a new catalytic motif. Org. Lett. 2010;12:5728–5731. DOI: 10.1021/ol102539x

[52] Zhou H, Zhang WZ, Liu CH, Qu JP, Lu XB. CO_2 adducts of *N*-heterocyclic carbenes: thermal stability and catalytic activity toward the coupling of CO_2 with epoxides. J. Org. Chem. 2008;73:8039–8044. DOI: 10.1021/jo801457r

[53] Kayaki Y, Yamamoto M, Ikariya T. *N*-Heterocyclic carbenes as efficient organocatalysts for CO_2 fixation reactions. Angew. Chem. Int. Ed. 2009;48:4194–4197. DOI: 10.1002/anie.200901399

[54] Naik PU, Petitjean L, Refes K, Picquet M, Plasseraud L. Imidazolium-2-carboxylate as an efficient, expeditious and eco-friendly organocatalyst for glycerol carbonate synthesis. Adv. Synth. Catal. 2009;351:1753–1756. DOI: 10.1002/adsc.200900280

[55] Wang YB, Wang YM, Zhang WZ, Lu XB. Fast CO_2 sequestration, activation, and catalytic transformation using *N* heterocyclic olefins. J. Am. Chem. Soc. 2013;135:11996–12003. DOI: 10.1021/ja405114e

[56] Wang YB, Sun DS, Zhou H, Zhang WZ, Lu XB. CO_2, COS and CS_2 adducts of *N*-heterocyclic olefins and their application as organocatalysts for carbon dioxide fixation. Green Chem. 2015;17:4009–4015. DOI: 10.1039/c5gc00948k

[57] Zhou H, Wang GX, Zhang WZ, Lu XB. CO_2 adducts of phosphorus ylides: highly active organocatalysts for carbon dioxide transformation. ACS Catal. 2015;5:6773–6779. DOI: 10.1021/acscatal.5b01409

[58] Gomes CDN, Jacquet O, Villiers C, Thuéry P, Ephritikhine M, Cantat T. A diagonal approach to chemical recycling of carbon dioxide: organocatalytic transformation for the reductive functionalization of CO_2. Angew. Chem. Int. Ed. 2012;51:187–190. DOI: 10.1002/anie.201105516

[59] Frogneux X, Blondiaux E, Thuéry P, Cantat T. Bridging amines with CO_2: organocatalyzed reduction of CO_2 to aminals. ACS Catal. 2015;5:3983–3987. DOI: 10.1021/acscatal.5b00734

Section 3

Photo-Redox and Electro-Organocatalysis

Chapter 6

Photo-Organocatalysis, Photo-Redox, and Electro-Organocatalysis Processes

Karolina Zalewska, Miguel M. Santos,
Hugo Cruz and Luis C. Branco

Additional information is available at the end of the chapter

http://dx.doi.org/10.5772/64633

Abstract

The research involving photo-organocatalysis, photoredox, and electro-organocatalysis processes is revised in this chapter. Modern synthetic processes enable the formation of large arrays of organic molecules with precise control over their three-dimensional structure, which is important in a variety of fields ranging from pharmaceutical to materials science. Photochemical reactions may have a substantial impact on these fields by affording direct access to specific structural motifs that are difficult to construct otherwise. The conjugate structural feature shown by most of the photo-organocatalysts seems to enable the production of free radicals or radical ions in an easy fashion. Electro-organocatalysis has also received recent interest from both academia and industry. In this chapter, we mainly review recent remarkable advancements in organocatalysis involving photo-, photoredox, and electrochemical processes with particular emphasis on asymmetric protocols.

Keywords: photochemistry, photoredox, electrochemistry, organocatalysis, sustainable chemistry

1. Introduction

In general, organocatalysts are divided into two main classes according to the interaction, covalent or non-covalent (H-bonding, proton transfer, ion pair formation), with the organic substrate within the catalytic cycle. In this context, an organocatalyst reacts with an organic molecule in order to form a stable organic compound or a labile intermediate. At this stage, the activation induced by the organocatalyst enables the attack of the second reagent to form a second adduct that releases the desired product with the concomitant regeneration of the organocatalyst.

Most of the common organocatalysts used for carbon-carbon bond formation reactions are based in chiral and achiral secondary amines, while reagents are electrophiles such as aldehydes, ketones, or α,β-unsaturated carbonyls. For these cases, the selected organocatalysts normally promote the generation of either an iminium ion or an enamine. *N*-heterocyclic carbenes have also been used as organocatalysts that promote the polarity inversion of an organic moiety for C—C bond formation.

Photocatalysis, where an electronically excited species acts as the catalyst, has gained increasing interest over the last years, with different organic transformations under such conditions being reported.

Recently, the catalytic activation of organic molecules by visible light photoredox catalysis that works under stereochemical control and provides chiral molecules in an asymmetric fashion has been largely reported. Generically, this approach relies on the ability of metal complexes and organic dyes to engage in single-electron transfer (SET) processes with organic substrates upon photoexcitation with visible light. Most common visible light photocatalysts are based on polypyridyl complexes of ruthenium, for example, tris(2,2′-bipyridine)ruthenium(II) or $[Ru(bpy)_3]^{2+}$, and iridium. These complexes absorb light in the visible region of the electromagnetic spectrum to give stable, long-living photoexcited states. The lifetime of the excited species is sufficiently long that it may engage in bimolecular electron transfer reactions in competition with deactivation pathways. Although these species are poor single-electron oxidants and reductants in the ground state, excitation of an electron affords excited states that are very potent single-electron transfer reagents. The ability of $[Ru(bpy)_3]^{2+}$ and related complexes to function as visible light photocatalysts has been recognized and currently applied to the electrolysis of water and the reduction in carbon dioxide to methane. These photocatalysts have also been employed in organic transformations including asymmetric approaches. Much of the excitement around visible light photoredox organocatalysis is due to the ability to achieve unique, if not exotic bond constructions that are not possible using the established protocols. For instance, photoredox organocatalysis can perform under overall redox neutral reactions where both oxidants and reductants are transiently generated in the same reaction vessel. This approach stands in contrast to methods requiring stoichiometric chemical oxidants and reductants, which are often incompatible with each other, as well as to electrochemical approaches, which are not amenable to redox neutral transformations.

Electro-organocatalysis has also received recent interest from both academia and industry. Electron transfer is one of the most important processes in organic chemistry in which one electron is added to or removed from an electroactive substrate. Such an electron transfer is reversible only when the resulting species are stable under those conditions. In other cases, an electron transfer generates subsequent chemical processes such as bond dissociation and bond formation. In general, radical cations and radical anions can be generated by electrochemical electron transfer reactions. Carbocations, carbon-free radicals, and carbanions can also be generated by subsequent bond dissociation or bond-forming processes. Several organic

http://dx.doi.org/10.5772/64633

synthetic transformations especially carbon-carbon bond formation reactions, oxidation, and reduction processes (electrocatalytic processes) have been reported.

2. Recent approaches in photo-organocatalysis

2.1. Asymmetric Photo-Organocatalysis

In the area of catalytic reactions, tremendous improvement has been made in the last decades, mostly upon the discovery of efficient transition metal catalysts. According to the variety of reactions, accessible, metal-catalyzed and enantioselective reactions have become significant tools in organic synthesis [1]. However, some disadvantages remain, such as the high cost and toxicity of the transition metal catalysts, employed and in some cases the problems that their residues, mainly in pharmaceutical products, can cause. Nonetheless, this transition metal catalysis will certainly continue to have an impact in synthetic organic chemistry in the future [2]. Alternatively, over the last years, a metal-free approach known as organocatalysis has reached a level of reliability that has allowed researchers to combine this procedure with other powerful techniques for molecule activation based on photochemical processes promoted by visible light. This green strategy has allowed previously unachievable synthetic issues to be solved and has rapidly progressed with application in both symmetric and asymmetric reactions (e.g., nucleophilic substitutions, Michael additions, cycloadditions, and aldol reactions) [3]. Generically, the organic catalysts can be categorized into two main classes according to the covalent or non-covalent (*viz.* H-bonding, proton transfer, ion pair formation) nature of the interaction established with the substrate within the catalytic cycle.

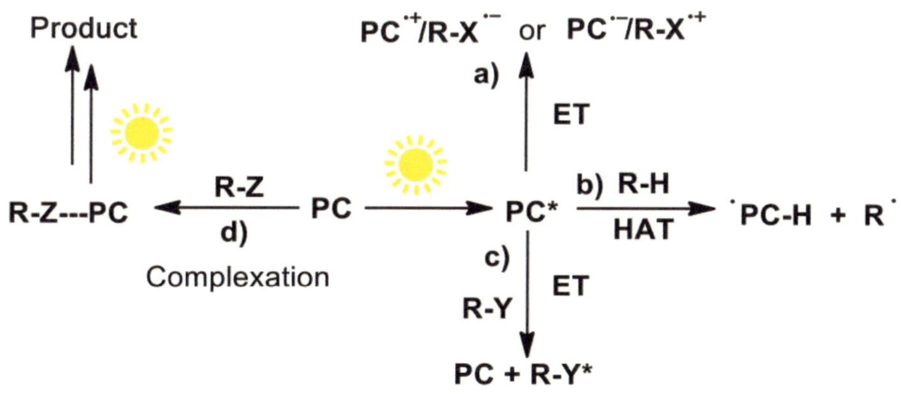

Scheme 1.

Homogeneous catalytic asymmetric transformations utilizing visible light photocatalysis include chiral and racemic photocatalysts with chiral organocatalysts, chiral Brønsted acids, or chiral Lewis acids [4]. In photo-organocatalytic processes, there are two main reaction models: the photocatalyst (PC) can act through an electron transfer (ET) process that causes an one-electron oxidation/reduction in the organic substrate R-X (**Scheme 1**, route a) or through hydrogen atom transfer (HAT, route b) from a hydrogen donor R-H [5]. Most of the photo-organocatalysts are aromatic ketones, dyes, and (chiral) secondary amines, while R substrates are electrophiles, typically aldehydes, ketones, or α,β-unsaturated carbonyls [5, 6]. Furthermore, photosensitization is known as an energy transfer between the excited photocatalyst (PC*) and substrate, which creates an excited state (R-Y*, from quenching of PC*), that is able to initiate a chemical reaction (route c). Sensitization can occurs by energy or electron transfer processes. The catalyst is transformed to act as a photosensitizer via photo-induced electron transfer (PET), hence leading the resulting photo-organocatalytic reaction to occur under stereoselective control [7].

Nowadays, a possible alternative can be considered in the photochemical activation step, in which the complexation of an organic reagent R-Z is controlled by a distinct, photostable chiral catalyst (route d) [8].

The aim of this subchapter was to point out the effective tools that the stereoselective ground-state processes offer to enantioselective photochemistry. The catalysts control the photoactivation of the substrates by inducing the transient formation of photon-absorbing chiral electron donor-acceptor (EDA) complexes. In addition, high stereocontrol in synthetically relevant intermolecular carbon-carbon bond-forming reactions driven by visible light can be provided by the inherent chirality of the catalysts.

The group of Bach focuses on catalytic processes, which allow previously unknown transformations employing both photochemical and conventional techniques. Their published papers concern photoredox organocatalysis, such as the first highly enantioselective (up to 90% *ee*) singlet oxygen [2+4] cycloaddition reactions [9], but also some related with the photo-organocatalysis. In 2005, Bach and co-workers presented an enantioselective photo-induced electron transfer (PET) sensitization with significant turnover and high enantioselectivity [10]. These PET-catalyzed conjugate additions of α-amino alkyl radicals to enones have already been studied non-enantioselectively [11]. For the first time, an electron-accepting chiral organocatalyst was applied, in contrast to conventional complexing reagents. The (pyrrolidinylethyl) quinolone (**PC1**; 30 mol%) that induces a chiral environment on the substrate through hydrogen bonding at the bridgehead lactam, lead to the formation of the spirocyclic pyrrolizidine product in high enantiomeric excess and yields (*ee* up to 70%, yield 64%).

Four years later, the same group tested the intramolecular [2+2] photocycloaddition of prochiral 4-(3′-butenyloxy) quinolone to the desired products (**Scheme 2**) [12]. The previously characterized chiral organocatalyst-benzophenone **PC1** indeed caused a rate acceleration of the photocycloaddition, but low stereoselectivity was achieved. In contrast, a novel synthesized xanthone **PC2** proved to be a more active catalyst resulting in significant rate acceleration by triplet energy transfer and high enantiomeric excess values. After initial optimization using 20 mol% of chiral organocatalyst, it was possible to obtain the products in 94% *ee* (**Table 1**, entry 12). Bearing in mind these two catalysts as a potential prototype for synthetically relevant transformations of quinolones, in 2011 Bach and co-workers reported the synthesis of six 2-quinolones and their use in intramolecular [2+2] photocycloaddition [13].

http://dx.doi.org/10.5772/64633

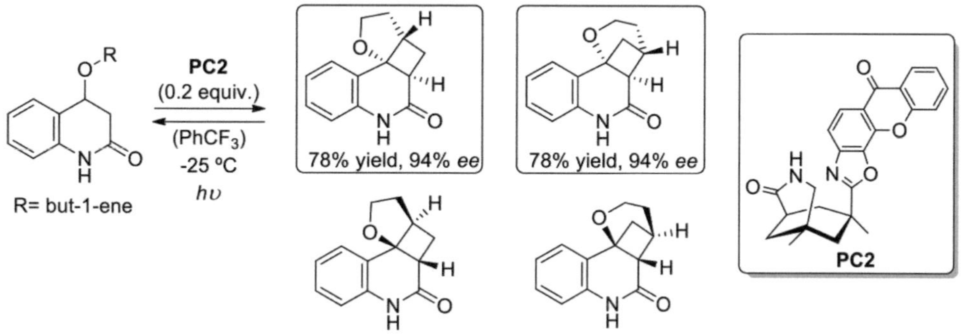

Scheme 2.

This photo-organocatalytic transformation was provided by applying a chiral, hydrogen-bonding template with an attached catalytically active sensitizing unit (benzophenone or xanthone). In all cases, it was possible to obtain high yields (78–99%) and enantioselectivities (83–94% *ee*) as shown in **Table 1**. These studies lead to a better understanding of stereoselective photo-organocatalytic processes, showing the importance of kinetic factors in creating an optimal catalytic cycle, as well as the activity range of different quinolones.

Entry	Substrate	λ (nm)	Catalyst (mol%)	Yield (%)[a]	*ee* (%)[b]
1	1	300	**PC3** 25	43	89
2	2	300	**PC3** 25	87	>90
3	3	300	**PC3** 25	66	83
4	4	366	**PC3** 25	89	89
5	5	366	**PC3** 25	99	90
6	6	366	**PC3** 25	99	94
7	4	366	**PC2** 10	58	92
8	4	366	**PC2** 10	75	90
9	4	366	**PC2** 10	50	91
10	4	366	**PC2** 10	46	89
11	4	366	**PC2** 5	48	90
12	4	366	**PC2** 20	53	94

[a] Yield of isolated product.
[b] The enantiomeric excess of the straight photocycloaddition products was determined by chiral HPLC analysis.

Table 1. Enantioselective photo-organocatalytic intramolecular [2+2]-photocycloaddition of quinolones.

In parallel, the group of Bach proposed an immobilization of earlier mentioned chiral photo-organocatalysts and their use in intramolecular [2+2] photocycloaddition of 4-allyloxyquinolone (**Scheme 3**) [14]. Under irradiation with light, the immobilized templates **PC4** and **PC5** allowed the substrate to undergo a [2+2] photocycloaddition to give the chiral products in high *ee* values and did not decrease even after the fourth use of recovered catalyst. Furthermore, the linking position of the catalyst at the C-6 carbon atom of the tetrahydronaphthalene was the one that rendered best results.

In different experiments, the group of Bach also investigated enantioselective photochemical reactions resorting on chiral Lewis acids as catalysts [15]. They reported the $AlBr_3$-activated chiral cationic oxazaborolidine catalyst for enantioselective intramolecular [2+2] photocycloaddition reactions of 4-alkenyl-substituted coumarins (78% *ee* were recorded with 20 mol % of catalyst). Nevertheless, the use of metals was inevitable.

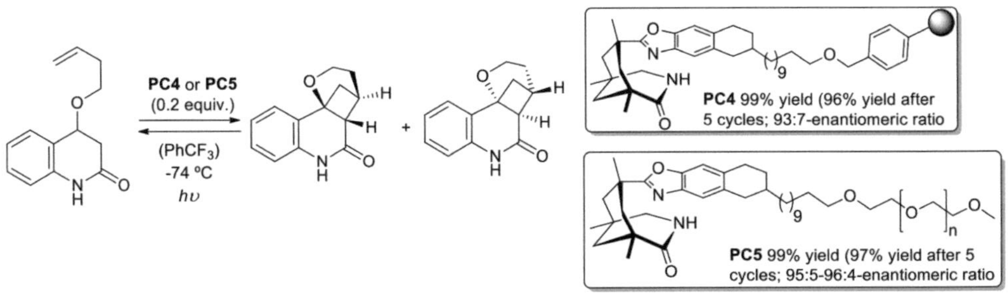

Scheme 3.

More recently, Vallavoju et al. [16] reported intramolecular [2+2] photocycloadditions of 4-alkenyl-substituted coumarins promoted by various atropisomeric binaphthyl-derived thioureas as photo-organocatalysts (**Scheme 4**). Thiourea catalysts are simple, environmentally benign, sustainable, and inexpensively synthesized from 'chiral pool', as well as easy to modulate and to handle. The photocatalytic cycle involves the formation of both static and dynamic complexes (exciplex formation) between the photo-organocatalyst and the reactive substrate, which are stabilized by hydrogen bonding. The corresponding products were achieved with high enantioselectivities (77–96% *ee* of product **1**, **Scheme 5**) with low catalyst loading (1–10 mol%). The authors tested the catalyst readily prepared in one step from commercially available, optically pure 2-amino-2′-hydroxy-1,1′-binaphthalene with different functional groups, in order to understand the interaction(s) between the catalyst and improve the stereoselectivity. It was discovered that using catalyst **PC6,** an excellent conversion and high enantioselectivity of the photoproduct would be obtained (84% conversion; 74% *ee* of product **1**). Nevertheless, the catalyst **PC8** showed a great potential as photo-organocatalyst for this transformation resulting in 100% product conversion and 96% *ee*. Additionally, it was proved that reducing the **PC8** catalyst loading from 100 to 30 to 10 mol% had a minimal impact on the enantioselectivity (94–96% *ee*; of product **1**) of the photoaddition product, and the reaction was completed in 30 min. By controlling the reactivity of the excited state through the formation of static and dynamic complexes, photocatalysts or sensitizers with higher excited-

http://dx.doi.org/10.5772/64633

state energies than the substrates can be completely avoided. This concept of catalysts may be a breakthrough in the discovery on new task-specific photo-organocatalysts.

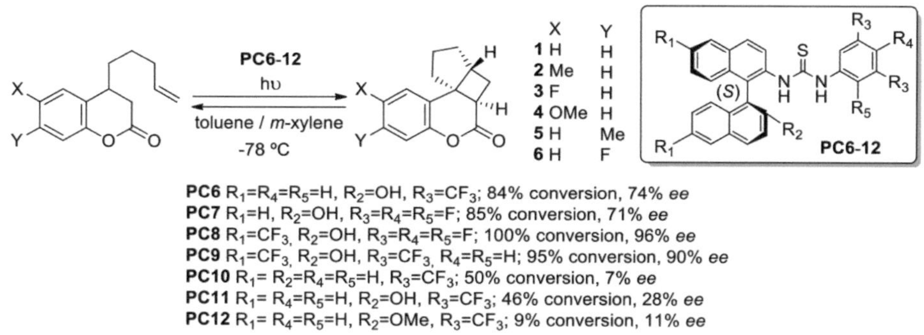

Scheme 4.

Melchiorre and co-workers presented the catalytic approach using a chiral organic catalyst with hydrogen-bonding motifs to bind a specific substrate selectively in synthetically relevant intermolecular carbon-carbon bond-forming reactions driven by visible light [7]. In the asymmetric α-alkylation of aldehydes with alkyl halides, the commercially available diarylprolinol silyl ether catalysts [17] **PC13** and **PC14** were chosen due to their ability to induce high enantioselectivity in thermal reactions of aldehydes that carry on through enamine formation. The authors tested the possibility of EDA complex formation by addition of an excess of butanal (15 equiv.) and amines **PC13** and **PC14** (1 equiv.) in methyl *tert*-butyl ether (MTBE) with 2,4-dinitrobenzyl bromide (1 equiv.). Using either 2,6-lutidine or sodium acetate as bases, this model reaction provided the desired α-benzylated product with high enantioselectivity (**Table 2**, entries 1 and 2, 83% *ee* with **PC14**). In its turn, the chiral organocatalyst **PC15** increased the *ee* value of the final product to 92%. The discovery of *in situ* created chiral EDA complexes from enamine intermediates which have the potential to participate actively in the photoexcitation of substrates without the required external photosensitizer, took the group of Melchiorre a step further.

In 2014, the authors describe the first light-driven enantioselective organocatalytic alkylation of unmodified ketones with alkyl halides [18]. This correlates to the previously established mechanism, in which the chiral enamines are the key intermediates in ground-state organocatalytic asymmetric processes. A variety of chiral primary amines (20 mol%) to activate cyclohexanone towards benzylation with 2,4-dinitrobenzyl bromide were studied. A chiral secondary amine did not show any ability to catalyze the photochemical alkylation; nevertheless, the primary amines displayed promising (entries 4 and 5) or even excellent (entry 6) reactivity, but insufficient enantioselectivity. The primary cinchona-based amine catalyst **PC16** confirmed the formation of photon-absorbing chiral electron donor-acceptor complexes, thus the photoactivation of the substrates. The benzylation product was obtained under cryogenic conditions (0°C) in a good yield and with an improved optical purity (60% yield, 90% *ee*, entry 7). Based on these optimized experiments, Melchiorre and co-workers examined different cyclic ketones in order to have an overview of the photochemical organocatalytic ketone alkylation approach. They discovered that a variety of N-Boc protected piperidine and dioxaspiro species can be readily active in this asymmetric alkylation reaction. Furthermore, the scope of the photochemical α-alkylation with diverse alkylating agents and cinchona-based amine catalyst led to the formation of α-alkylated products with high levels of regio-, diastereo-, and enantioselectivity.

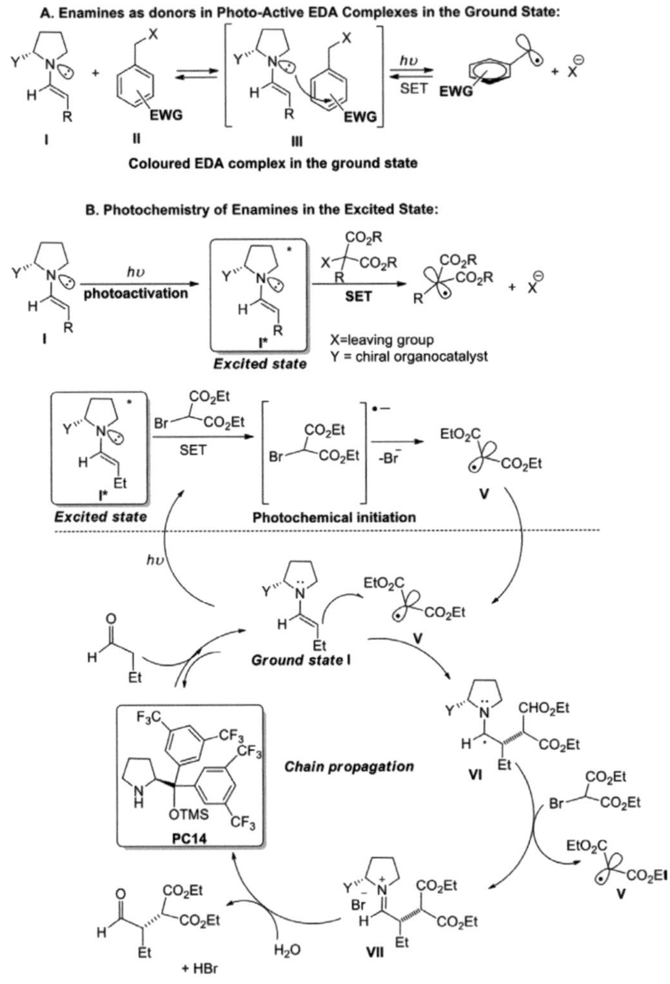

Scheme 5.

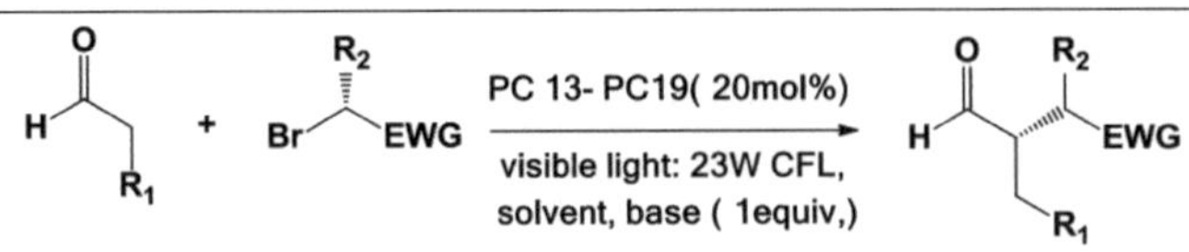

http://dx.doi.org/10.5772/64633

Entry	Ketone or aldehyde	Alkyl halide substituted with EWG	Catalyst	Yield (%)	*ee* (%)	Ref.
1	-$(CH_2)_2$-CH_3	2,4-dinitrobenzyl bromide	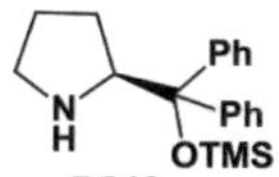 PC13	98	75	[7]
2			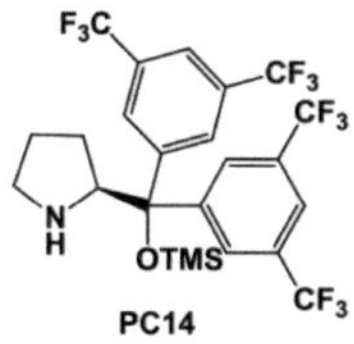 PC14	98	83	[7]
3			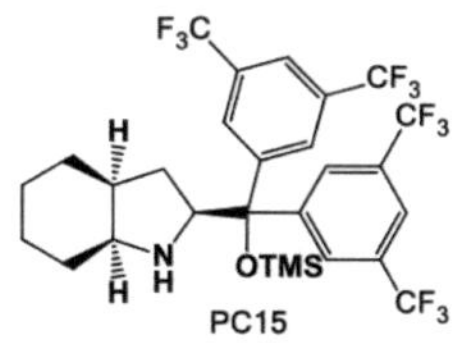 PC15	87	92	[7]
4	Cyclohexanone		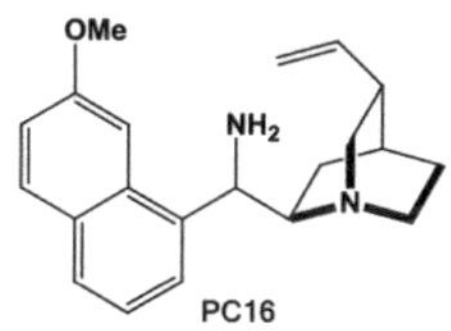 PC16	45	88	[18]
5		2,4-dinitrobenzyl bromide	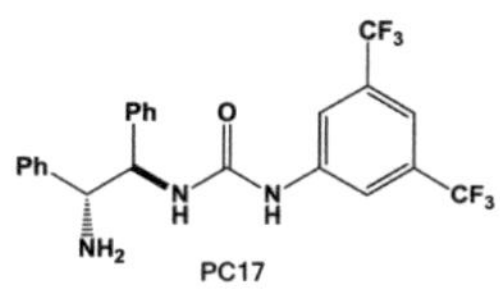 PC17	45	42	[18]
6			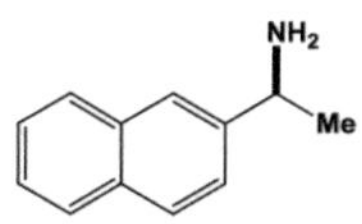 PC18	90	18	[18]
7			PC16	60	90	[18]
8			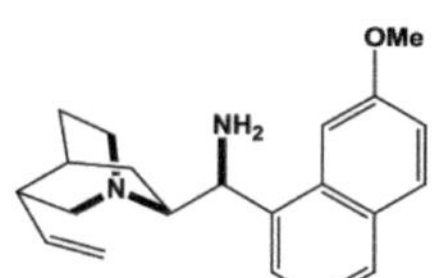 PC19	50	90	[18]
9	4,4-dimethylcyclohexanone		PC16	65	82	[18]
10	3,3-dimethylcyclohexanone		PC16	57	94	[18]
11	4-methylcyclohexanone		PC16	94	94	[18]
12	1,4-dioxaspiro[4.5]decan-8-one		PC16	70	95	[18]
13	-$(CH_2)_2$-CH_3	Diethyl-2-bromomalonate	PC14	94	83	[19]

Entry	Ketone or aldehyde	Alkyl halide substituted with EWG	Catalyst	Yield (%)	*ee* (%)	Ref.
14	-$(CH_2)_2$-CH_3	diethyl 2-bromo-2 -methylmalonate	PC14	98	91	[19]
15	-$(CH_2)_4$-TIPSO	diethyl 2-bromo-2 -methylmalonate	PC14	94	88	[19]

Table 2. Asymmetric α-alkylation of aldehydes and ketones with alkyl halides by photo-organocatalysis.

Last year, the group of Melchiorre reported the photo-organocatalytic enantioselective α- and γ-alkylation of aldehydes and enals with bromomalonates by a fluorescent light bulb without the need of any external photoredox catalyst [19]. The preliminary studies involved butanal and diethyl bromomalonate (**Table 2**) as substrates for this photo-organocatalytic reaction. The results showed that using the aminocatalyst **PC14** (20 mol%) in a MTBE solution under irradiation, the alkylation product was obtained in high yield and enantioselectivity after 4 h (94% yield, 83% *ee*; entry 13). The detailed photochemical studies using absorption and emission spectroscopy suggested that the direct photoexcitation of the enamine could trigger the radical generation from diethyl bromomalonate. Furthermore, no photoabsorbing ground-state EDA complex formation was observed. Accordingly, the photochemical reaction proceeds through a different mechanism. As previously described, the metal-free process depends on the creation of photon-absorbing electron donor-acceptor (EDA) complexes [18], generated in the ground state upon association of electron-deficient benzyl and phenacyl bromides II with the electron-rich enamine I (**Scheme 5A**). A single-electron transfer (SET) induced by visible light irradiation of the colored EDA complex III allows access to radical species under mild conditions. This reactivity allowed the expansion of a light-driven stereo-selective α-alkylation of carbonyl compounds [18]. In contrast to this, the authors suggest a novel photo-organocatalytic mechanism, in which enamines can use to drive the photochemical generation of radicals acting as a photosensitizer upon direct photoexcitation (**Scheme 5B**). The enamine **I**, under light absorption, reaches an electronically excited state (**I***) and to act as a photoinitiator causing the formation of the electron-deficient radical **V** through the reductive cleavage of the bromomalonate C—Br bond via a single-electron transfer (SET) mechanism (**Scheme 5B**) [19, 20]. Taking into consideration the capacity of **I** to infer high stereoselectivity in enamine-mediated polar reactions, adding the radical **V** to the ground-state **I** progresses in a stereocontrolled manner. Since α-aminoalkyl radicals known as strong reducing agents, the intermediate **VI** would induce the reductive cleavage of bromomalonate through an outer-sphere SET process, thus regenerating the radical **V**. This route affords a bromide iminium ion pair **VII**, which then hydrolyzes to release the desired product and the aminocatalyst **PC14**. Then, the scope of enantioselective organocatalytic alkylation of aldehydes and enals was examinated. Differently substituted bromomalonates successfully contributed to the enantio-selective alkylation of butanal at room temperature (**Table 2**, entry 14). Aldehydes with a heteroatom moiety, long-alkyl chain, or an internal olefin, were also stereoselectively alkylated to give products with good enantioselectivity (**Table 2**, entry 15). Additionally, it was pre-

http://dx.doi.org/10.5772/64633

sented that enamine intermediates, synthesized from α-branched enals, could trap the generated radical while locating a new stereocenter at a distant γ-position.

In parallel, the same group of researchers investigated the phase transfer catalyzed, enantioselective perfluoroalkylation and trifluoromethylation of cyclic β-ketoesters under visible light irradiation [21]. The photo-organocatalytic approach is again caused by the photochemical activity of EDA complexes generated *in situ* from the ground-state association of chiral enolates and perfluoroalkyl iodides. Irradiating the colored EDA complex induces a single-electron transfer (SET) allowing access to radical species at ambient temperature. Perfluoroalkyl iodides were selected as electrophiles due to their potential for facilitating EDA associations in the ground state. At the same time, chiral quaternary ammonium salts were used as phase transfer catalysts (PTC) that enabled the formation of a chiral ion pair after deprotonation of β-ketoesters by an inorganic base [22]. Perfluoroalkylation of one indanone methyl ester with perfluorohexyl iodide in chlorobenzene under visible light irradiation with PTC organocatalyst was chosen as model reaction. The best results were obtained using the *pseudo*-enantiomeric cinchonine **PC20** derivative catalysts, with an excess of perfluorohexyl iodide (3 equiv) for 64 h (59% yield, 93% *ee*) (**Scheme 6**) [21].

The above presented strategies of enantioselective photo-organocatalytic processes have a great potential for the sustainable preparation of chiral molecules, a rapidly developing area of modern chemical research.

In parallel to the efforts performed in the field of asymmetric photo-organocatalysis, some attempts were also performed in the non-enantioselective processes.

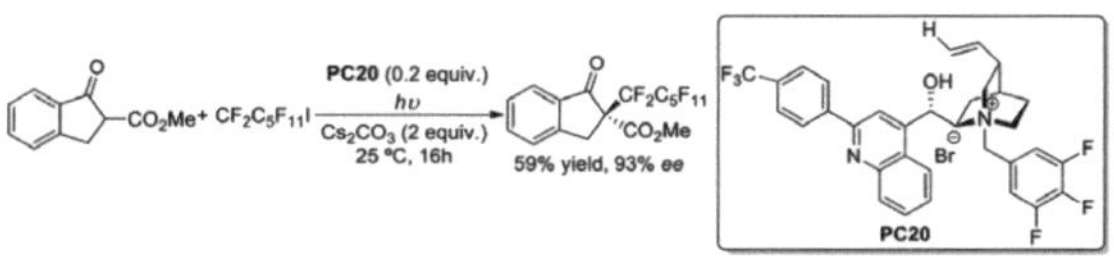

Scheme 6.

Non-asymmetric photocatalysis has gained a great deal of attention during the last decades [23, 24], and a remarkable and interesting case was recently described by the already cited group of Melchiorre, in which an aromatic aldehyde was involved in the intermolecular atom transfer radical additions (ATRA) of a variety of haloalkanes to alkenes, one of the essential carbon-carbon bond-forming processes in organic chemistry [25]. In an ATRA reaction, the addition of an organic halide across a carbon-carbon double-bond yields a new C—C and C—X bond (X = halogen) in a single operation. Once more, organic compounds known to be capable of high photoreactivity [25] could alternatively be used as an energy transfer photocatalyst. It is important to note that for the first time, aromatic aldehydes have been used as photo-organocatalysts in an effective and valuable process [26]. Recent exciting findings by Melchiorre and co-workers have also shown the metal-free photo-organocatalysis which allows the direct alkylation of 2- and 3-substituted 1*H*-indoles with electron-accepting benzyl and phenacyl bromides [27].

3. Recent approaches in photoredox organocatalysis

The term photoredox organocatalysis has its origin in the work by Nicewicz and MacMillan in 2008. They reported the enantioselective α-alkylation of aldehydes using $[Ru(bpy)_3]Cl_2$ as a photoredox catalyst. This complex, alongside many others such as $[Ir(ppy)_2(dtb\text{-}bpy)]PF_6$ and *fac*-$[Ir(ppy)_3]$ that have been subsequently reported, acts as strong oxidizers in the excited state upon absorbing visible light. In general, different inorganic solid photocatalysts such as TiO_2 and ZnO have been largely explored as photoredox catalysts. The poor absorption of visible light by such inorganic photocatalysts is considered a limitation for application in organic synthetic processes using solar energies. In contrast, organic photocatalysts show some advantages regarding their low-cost, significant synthetic versatility, and the possibility to tune their redox properties. In this context, the organic photoredox dyes are usually selected according to their λ_{max} and redox potential E^0.

Scheme 7 depicts the most popular dyes investigated in photoredox catalysis procedures.

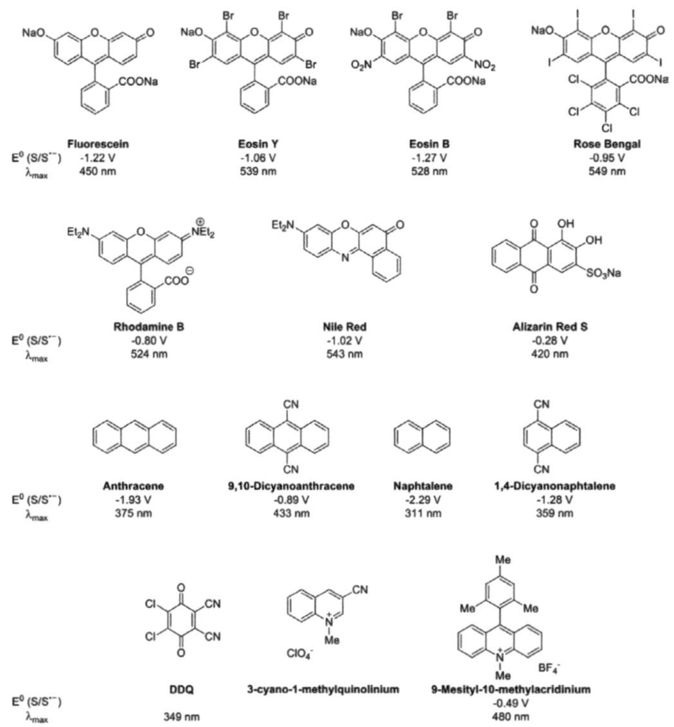

Scheme 7.

The photoactivation reveals the ability of the photosensitizer to absorb in the visible domain and to act both as a strong oxidant in the excited state S* and as an efficient reductant in its semi-reduced form $S^{\bullet-}$. In **Scheme 8**, a comparison between the general photoredox catalytic cycles of ruthenium-based catalysts and a photo-organocatalyst, *viz*. Eosin Y (EY), is presented.

http://dx.doi.org/10.5772/64633

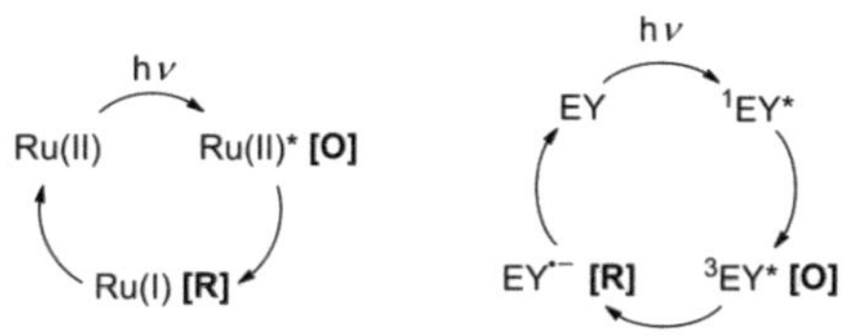

Scheme 8.

3.1. Asymmetric photoredox organocatalysis

One of the most explored aspects investigated in the field of enantioselective photoredox catalysis has been the use of organic dyes as photocatalysts. In the seminal work by Zeitler *et al.* [28], they reported the efficient cooperative asymmetric intermolecular α-alkylation of aldehydes catalyzed by Eosin Y under LED green light and in the presence of MacMillan's imidazolidinone catalyst **A** (see **Scheme 9** for a simplified proposed mechanism of the reaction). According to the electron-withdrawing groups of the substrates [diethylmalonate, *p*-nitrophenyl (PNP) and polyfluorinated alkane], the reactions performed better between −15 and +5°C with yields from 56 to 85% and enantiomeric excesses higher than 86%, despite the 18 h reaction time (see **Table 3**, entries 1–4).

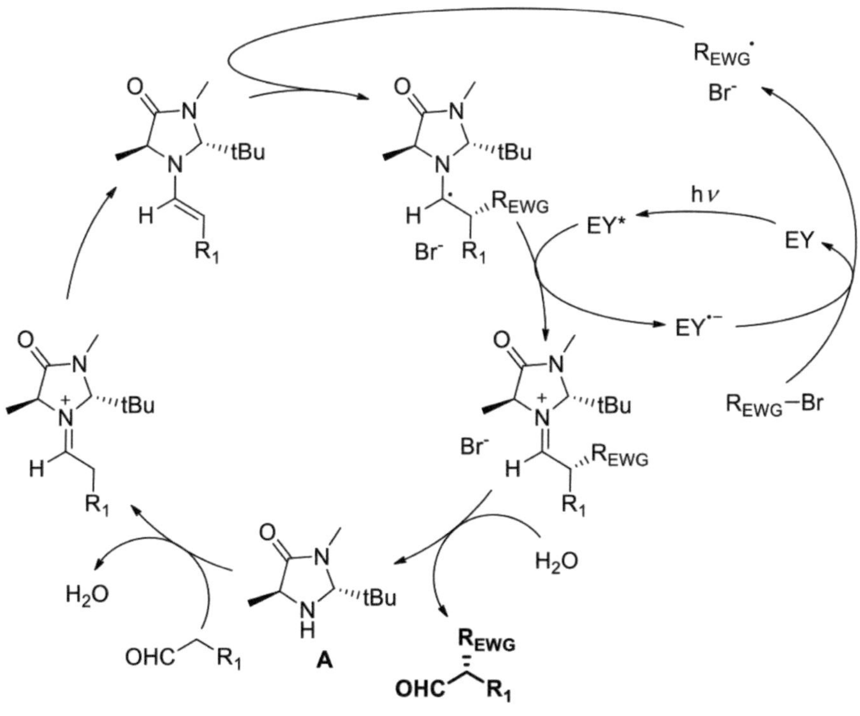

Scheme 9.

Using the same conditions adapted to a microreactor flow regime, smaller reaction times were obtained with comparable results [29]. Rose Bengal was also applied as photoredox catalyst in this type of reaction (**Table 3**, entries 3–9). Again, imidazolinone **A** was employed as well as a Lewis acid such as LiCl to co-catalyze the photoreaction [30].

On the other hand, asymmetric α-amination of aldehydes has also been accomplished by means of photoredox chemistry [31]. By using an amine substrate bearing ODNs, photolabile groups that simultaneously work as the photoredox catalyst and also release the reactive carbamyl reagent that couples with the *in situ* formed enamines; sixteen α-amino aldehydes were successfully prepared in 67–79% yield and >86% *ee* (see **Table 4**).

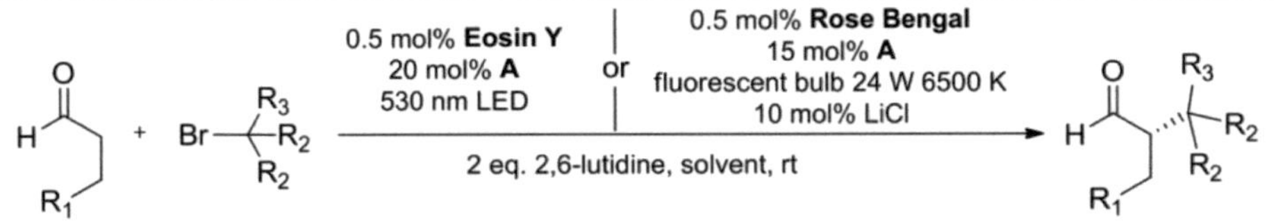

Entry	R₁	R₂	R₃	Yield (%)	*ee* (%)	Ref.
1	*n*-Pentane	H	$(CF_2)_3CF_3$	56	96	[28]
2	*n*-Pentane	H	PNP-CO	82	95	[28]
3	Ph	CO_2Et	H	76	86	[28]
	Ph	CO_2Et	H	89	83	[30]
4	*n*-Pentane	CO_2Et	H	85	88	[28]
	n-Pentane	CO_2Et	H	88	80	[30]
5	4-*t*BuPh	CO_2Et	H	51	83	[30]
6	3-ClPh	CO_2Et	H	90	82	[30]
7	4-OMePh	CO_2Et	H	94	80	[30]
8	2,4-$(OMe)_2$Ph	CO_2Et	H	51	83	[30]
9	$(CH_2)_2CH{=}CHEt$	CO_2Et	H	56	85	[30]

Table 3. Asymmetric alkylation of aldehydes catalyzed by Eosin Y or Rose Bengal.

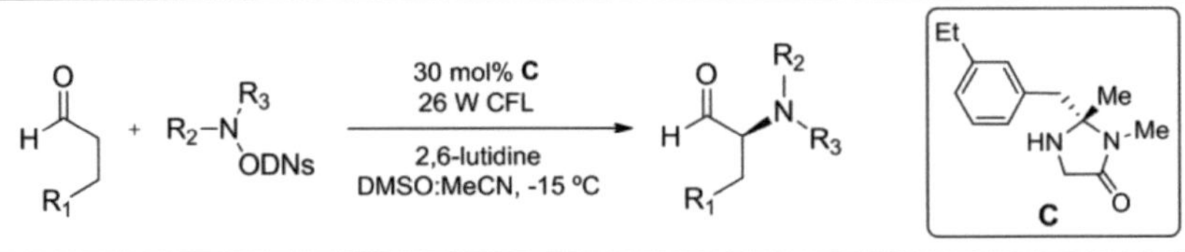

Entry	R_1	R_2	R_3	Yield (%)	*ee* (%)	Ref.
1	Ph	Me	Cbz	77	89	[31]
2	Ph	Me	Alloc	75	90	[31]
3	Ph	Me	Boc	71	89	[31]
4	Ph	Me	Fmoc	73	89	[31]
5	Ph	Bu	CO_2Me	76	90	[31]
6	Ph	MOM	CO_2Me	75	94	[31]
7	Ph	$(CH_2)_3Ph$	CO_2Me	71	86	[31]
8	Ph	Me	CO_2Me	79	92	[31]
9	*n*-pent	Me	CO_2Me	71	90	[31]
10	$(CH_2)_2OBn$	Me	CO_2Me	73	88	[31]
11	$(CH_2)_2ONPhth$	Me	CO_2Me	77	90	[31]
12	$CH{=}CH_2$	Me	CO_2Me	76	90	[31]
13	$(CH_2)_2CO_2Et$	Me	CO_2Me	71	90	[31]
14	PMP	Me	CO_2Me	79	91	[31]
15	Cyclohexyl	Me	CO_2Me	72	91	[31]
16	Me_2	Me	CO_2Me	67	94	[31]

Table 4. Asymmetric α-amination of aldehydes by ODNs dual catalysis.

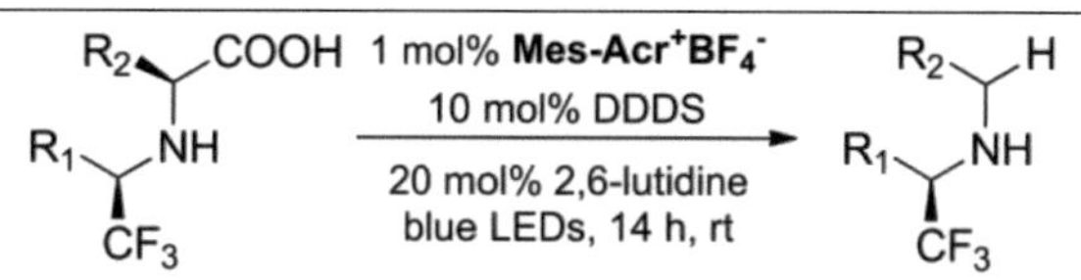

Entry	R_1	R_2	Yield (%)	*ee*(%)	Ref.
1	4-FPh	*i*Pr	72	83	[32]
2	Ph	*i*Pr	75	97	[32]
3	Ph	$(CH_2)_2SCH_3$	76	69	[32]
4	Ph	Bn	37	88	[32]

Table 5. Decarboxylative reduction in 1-aryl-2,2,2-trifluoroethyl-substituted amino acids.

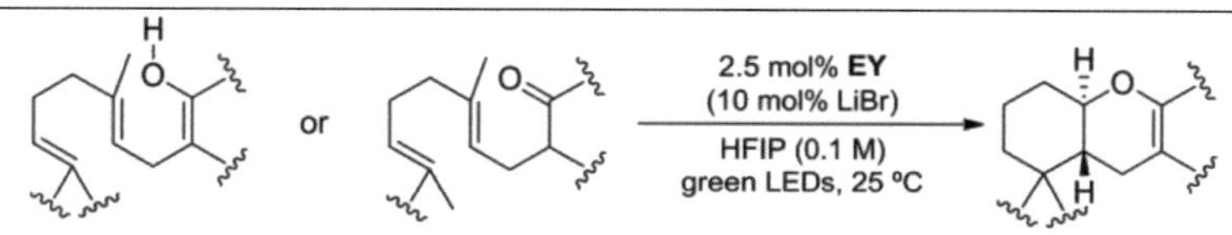

Entry	Compounds	R	Yield (%)	d.r. (%)	Ref.
1		Br	93	> 19:1	[33]
2		CO_2Me	79	> 19:1	[33]
3		F	85	> 19:1	[33]
4	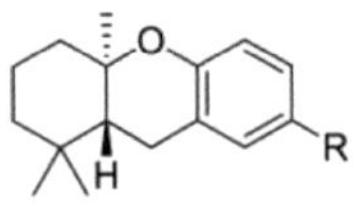	Cl	85	> 19:1	[33]
5		OCF_3	83	> 19:1	[33]
6		I	83	> 19:1	[33]
7		OMe	83		
8		Br	72	> 19:1	[33]
9	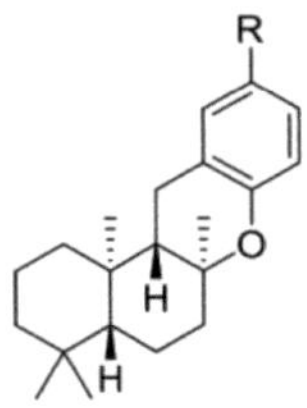	Cl	40	> 19:1	[33]
10	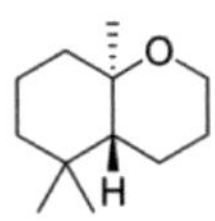	–	70	> 19:1	[33]
11	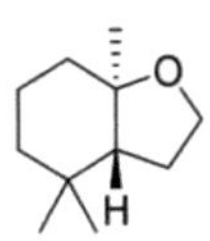	–	65	> 19:1	[33]
12		Ph	Ph	88	[33]
13		Ph	OEt	88	[33]
14	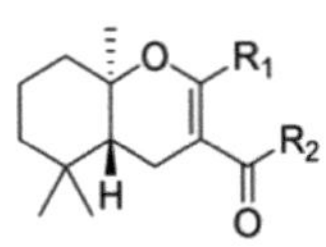	Ph	Me	87	[33]
15		Me	OEt	91	[33]
16	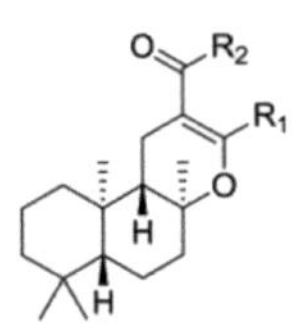	Ph	OEt	60	[33]

Table 6. Cyclization of polyprenoids and 1,3-ketocabonyls catalyzed by Eosin Y.

Wallentin et al. [32] reported the photocatalyzed decarboxylative reduction in several classes of biologically relevant enantio-enriched 1-aryl-2,2,2-trifluoroethyl-substituted amino acids (see **Table 5**). A plausible redox-coupled hydrogen shuttle mechanism was proposed by using one of the strongest oxidizing organic dyes mesityl acridinium (Mes-$Acr^+BF_4^-$,

$E_{1/2}^{red}$ = +2.06 V *vs* SCE) as photoredox catalyst and bis(4-chlorophenyl)disulphide (DDDS) as a sacrificial hydrogen atom donor. This methodology was also applied to the synthesis of other achiral carboxylic acids, namely α-amino acids, α-hydroxy acids, and phenylacetic acids in moderate to quantitative yields.

A stereoselective radical cascade cyclization of polyprenoids through a photocatalytic mechanism has been reported yielding polyenes in moderate to very high yields with excellent diastereoselectivities (d.r. > 19:1) in HFIP and using Eosin Y as the photoredox catalyst (**Table 6**) [33]. The methodology was based on the cyclization by terminal OH groups of a large substrate array of aliphatic alcohols, phenols, or enols, which was tolerable to electron-rich or electron-poor substituents. In addition, the cyclization of 1,3-diketones required the use of LiBr as a weak Lewis acid. Stern-Volmer analysis reinforced that these reactions proceeded *via* a PET-induced radical mechanism.

The photoredox catalyst 2,4,6-*tris*(4-methoxyphenyl)pyrylium tetrafluoroborate (**D**) was used in the enantioselective ring opening metathesis polymerization (ROMP) of *endo*-DCPD to provide the corresponding linear polymer with conversion yields as high as 20% [34] (**Scheme 10**). The copolymerization of this monomer with norbornene was also accomplished. By using partially hydrogenated monomers, side reactions at the terminal positions were hindered which enabled higher degrees of polymerization to be obtained. By applying visible light, the polymerization can be turned on or off at will, bringing a dynamic nature to it by exposure to the stimulus.

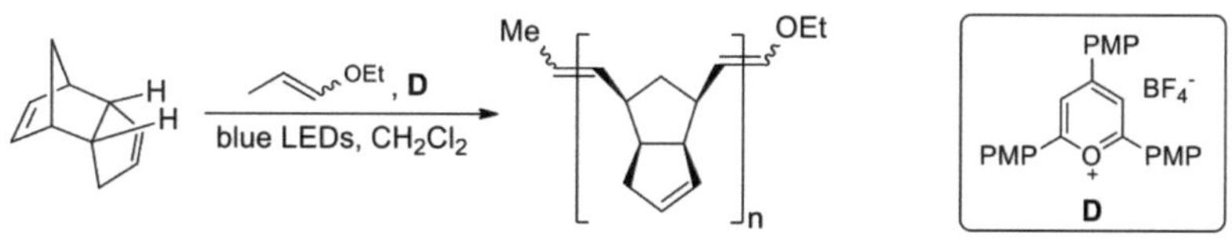

Scheme 10.

Over the last years, several publications involving non-asymmetric photoredox organocatalytic synthetic transformations mediated by metal-free organic photoredox catalysis under mild conditions have been reported [35].

From the industrial point of view, it is important to focus the recent developments on selective photocatalytic transformations of benzene, in particular the oxidation of benzene to phenol [36], alkoxylation of benzene [37], and monofluorination of benzene with fluoride and oxygen [38]. As an alternative to inorganic catalysts, the selective oxidation of benzene to phenol can be made under visible light irradiation of 2,3-dichloro-5,6-dicyano-*p*-benzoquinone (DDQ) in an oxygen-saturated acetonitrile solution of benzene and water [39]. In 2004, Fukuzumi *et al.* [40] reported the use of 9-mesityl-10-methylacridinium (Acr^+-Mes) ion as an efficient photoredox catalyst due to its high oxidizing and reducing abilities of the long-lived electron transfer (ET) state. Acr^+-Mes provides an efficient mediator to the formation of radical cations of electron donors and radical anions of electron acceptors thus enabling coupling between electron donors and acceptors.

Photocatalytic [2+2] cycloaddition of dioxygen to tetraphenylethylene (TPE) *via* electron transfer reactions of TPE and oxygen with the ET state of Acr^{+}-Mes is one example of the strategies described before [41].

Eosin Y as a well-known low-cost organic dye that absorbs green light (characteristic peak at 539 nm) has been extensively investigated as photoredox catalyst for different organic transformations [42–49].

4. Recent approaches in electro-organocatalysis

As described in the previous sections, the electron transfer (ET) process is a crucial step in the organic chemistry field on which many organic reactions rely in order to occur [50, 51]. Essentially, an electron transfer process is based on the removal (or addition) of at least one electron from (or to) the electroactive substrate. This process is considered reversible only when the obtained products are stable under those experimental conditions. An electron transfer can generate intermediates which subsequently undergo chemical processes such as bond dissociation and bond formation. Basically, electroctrochemical techniques can be applied to establish the electrochemical oxidation and reduction mechanisms, that is the electron transfer reaction (formation and determination of the intermediates) and subsequent chemical reaction associated with the electrochemical generated process (formation of the reaction products). Thus, those formed intermediates are radical cations (or radical anions), and they can be generated by electrosynthetic processes using organic compounds. Carbon-free radicals (carbocations and carbanions) can also be generated by subsequent bond dissociation or bond formation process. Several electro-organic synthetic transformations, especially carbon-carbon, carbon-nitrogen, and carbon-phosphorous bond formation reactions, as well as oxidation and reduction processes have been reported [52]. Electrochemical processes are considered 'green' procedures for those synthetic transformations. The main advantage of the electrosynthetic approach is that electrons flow as current and are regard as one inexpensive reactant, thus making the route more environmentally friendly. Moreover, reactions take place in low-temperature conditions, reducing the local consumption of energy and the risk of corrosion, material failure, and accidental release. Finally, it is important to highlight that electrodes can be regarded as heterogeneous catalysts that are easily separated from the products. The low or even almost inexistent volatility of the reaction media is another factor to be taken into account. Therefore, electro-organocatalysis constitutes a valuable tool for the organic chemist with numerous applications in both academia [53, 54] and industry [55].

The electro-organocatalysis field can be divided into two main branches, depicted in **Scheme 11(A)** direct electrolysis, in which the redox process occurs between the electrode surface and the reactant without the addition of other compounds and (B) 'indirect electrolysis' where the redox process occurs between the electrode surface and an external redox catalyst (or 'mediator') which then performs the ET with the reactive species) [56].

http://dx.doi.org/10.5772/64633

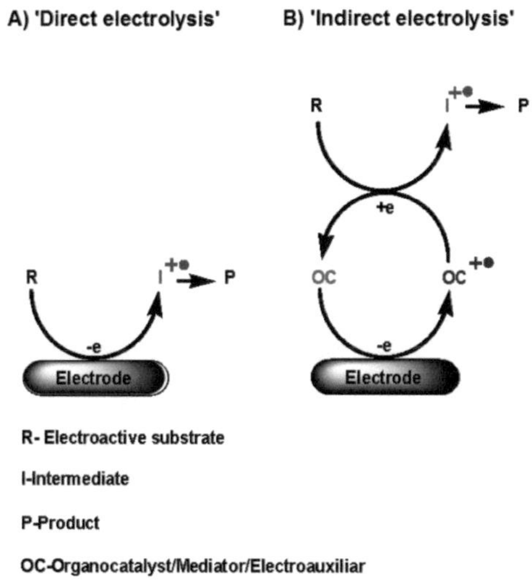

Scheme 11.

4.1. Direct electro-organocatalysis'

In the direct electro-organocatalysis process, the electron transfer (ET) step occurs at the electrode surface. Due to its heterogeneous nature, the catalyst recycling can be performed easily by separating it from the reaction media after the formation of the desired organic product.

'Direct electro-organocatalysis' or electro-organic synthesis has recently gained increasing attention, which can be attributed to their sustainable and 'green' features when compared to the traditional ones.

In the literature, there are few reports concerning bond formation and bond dissociation reactions. Gallardo and co-workers reported the formation of C—C [52, 57, 58], C—N [58], C—P [59], and C—S [60] bonds by an electrochemical approach of nucleophilic aromatic substitution reactions (S_NAr). The proposed new route for the electrochemical processes consists on the reaction between an electron-deficient, aromatic compound and a nucleophile, leading to the formation of a σ-complex or Meisenheimer complex intermediate. Then, this species undergoes an oxidation that leads to the departure of the leaving group (heteroatom radical [NASX] and/or hydride, two electrons and a proton [NASH]). This procedure was similarly conducted with other nucleophiles (hydride, cyanide, fluoride, methoxy, ethanethiolate, and n-butylamine) and aromatic compounds as starting materials. In addition, preparative electrolysis was also employed as means to promote the oxidation of the intermediate produced in the first step of the process [52, 58].

This technique allows determination, characterization, and quantification of the type and number of electrochemically produced complexes present in the reaction media. It is also possible to assess if the reaction was successful once most classical S_NAr reactions give lower yields.

The main drawbacks of the electrochemical approach are the use of solvent and the amount of tetraalkylammonium salt as electrolyte, which consequently have to be separated from the desired product. The use of ionic liquids (ILs) in particular room temperature ionic liquid (RTIL) as solvents may address this specific problem. They are considered non-flammable, non-volatile, and thermally stable over a wide range of temperatures, as well as good solvents for organic and inorganic compounds. In addition, they may be applied concomitantly as solvent and as electrolyte thereby enhancing the 'green' aspect of these procedures.

Gallardo and co-workers [61] adapted the electrochemical approach of nucleophilic aromatic substitution reactions to this 'greener' alternative family of solvents. The authors described the investigation of the electrocatalytic process as well as regioselectivity effects induced by the solvation properties of the RTILs (1-butyl-3-methylimidazolium [BMIM] combined with tetrafluoroborate [BF_4], hexafluorophosphate [PF_6], bis(trifluoromethylsulfonyl)imide [NTf_2], and acetate [AcO] as anions).

The use of electrochemical techniques such as cyclic voltammetry (CV) and controlled potential electrolysis allows the evaluation of the nature and stability of the electrochemically generated intermediate on the solvent, as well as the extension of the reaction.

Despite the successful reports on S_NAr reactions, the 'direct electrolysis' approach requires the application of high potentials in order for the electrosynthetic process to occur. To address this issue, the redox process can be applied to organocatalysts which then lead to yield the desired products in the indirect electrocatalysis fashion.

4.2. Indirect electro-organocatalysis

In the indirect electro-organocatalysis process, the electron transfer (ET) step is shifted from a heterogeneous process occurring at the electrode surface (as described earlier as 'direct electrolysis') to homogeneous process that can provide an electrochemically generated substance which acts as a so-called organocatalyst (or 'mediator'). Usually triarylamines, triarylimidazoles and *N*-oxyl radicals [62] are employed as these electroauxiliary species. The group of R. D. Little has reported several environmentally friendly methodologies to obtain products *via* 'indirect electro-organocatalysis' in which no metal catalyst and external chemical oxidants were employed [63–65].

In order to explore and generalize this methodology, analogous organocatalysts with modified aromatic rings were also reported by the authors. The desired products were formed in good yields [63].

4.3. Asymmetric electro-organocatalysis

In this specific case of the 'indirect electro-organocatalysis', particular conditions of solvent and catalyst are employed in order to enhance the enantioselectivity of the formed products. It is considered as safer and 'green route' towards enantioselective reactions by combining asymmetric organocatalysis with electrochemistry. The selected organocatalysts are stable, stereoselective organic compounds that can undergo the electrosynthetic process under unsuitable conditions for conventional catalysts. Asymmetric electro-organocatalysis methodologies have been successfully employed to produce several optically active compounds with application in life sciences. **Scheme 12** depicts the direct intermolecular α-arylation of aldehydes to produce meta-alkylated anilines using electron-rich aromatic compounds [66].

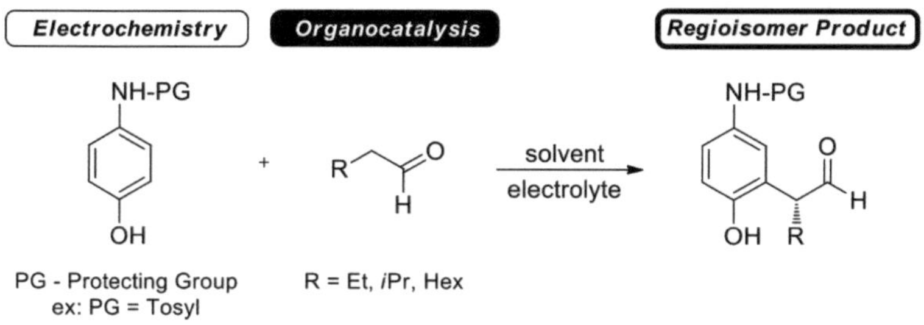

Scheme 12.

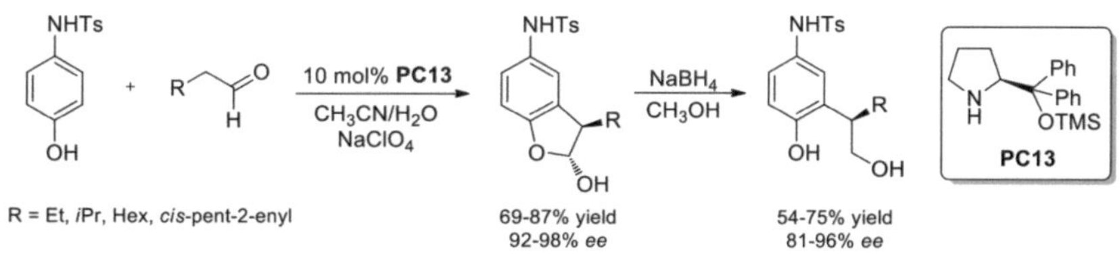

Scheme 13.

The described methodology for the regio- and stereoselective electroorganocatalyzed production of the *meta*-substituted anilines takes place in two steps: (1) firstly occurs the electrochemical activation of the aromatic compound that evolves to an electrophilic intermediate and (2) then an electron-rich enamine is generated by the condensation of the organocatalyst and the desired aldehyde, which subsequently undergoes a nucleophilic addition to the intermediate formed in step 1, giving rise to another intermediate that upon hydrolysis and proton transfer regenerates the organocatalyst and yields the corresponding product. The described reactions occur between the tosyl-protected reactant with a series of aldehydes catalyzed by [(*S*)-2-[diphenyl(trimethylsilyloxy)methyl]pyrrolidine] (**PC13**), which has also been reported for the photocatalyzed α-alkylation of aldehydes [7]. In these conditions, the *meta*-substituted aniline enantiomeric products were obtained in 54–75% yields and high enantiomeric excess between 81 and 96% (**Scheme 13**) [66].

In 2005, Schäfer and co-workers [67] reported the reaction of enamines and mediated anodic oxidation of carbohydrates in the presence of 2,2,6,6-tetramethylpiperidine-1-oxoammonium cation ([TEMPO]) as organocatalyst. These species reacted with selected enaminoesters to form intermediate imidazolium cations, which selectively oxidize the primary hydroxy groups of trisaccharides at the anode to give tricarboxylic acid sugars in 50–80% yields. The relative stability of the electrogenerated TEMPO cation in acetonitrile enables it to react as a selective oxidant, electrophile, and also catalyst.

Enantioselective α-oxyamination of aldehydes has been reported by the group of H.-J. Jang using a *sec*-amine as chiral catalyst (**Scheme 14**) [68]. An enamine intermediate is formed by anodic oxidation of the aldehydes that ultimately reacts with TEMPO leading to the formation of the desired products in reasonable yields. Once again using **PC13** as organocatalyst, α-oxyaldehydes were obtained in 23–57% yields and 60–70% *ee*.

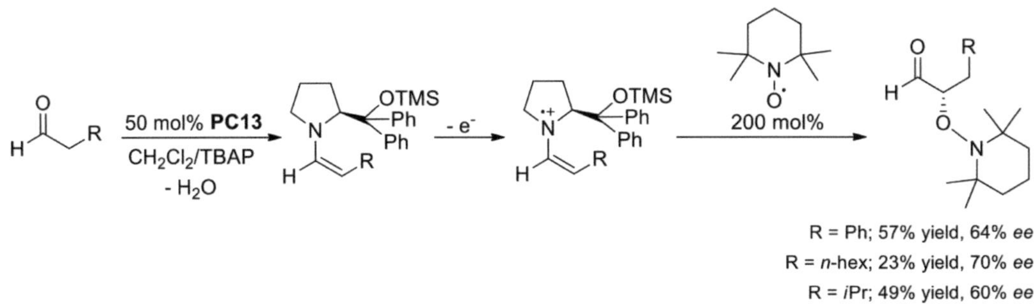

Scheme 14.

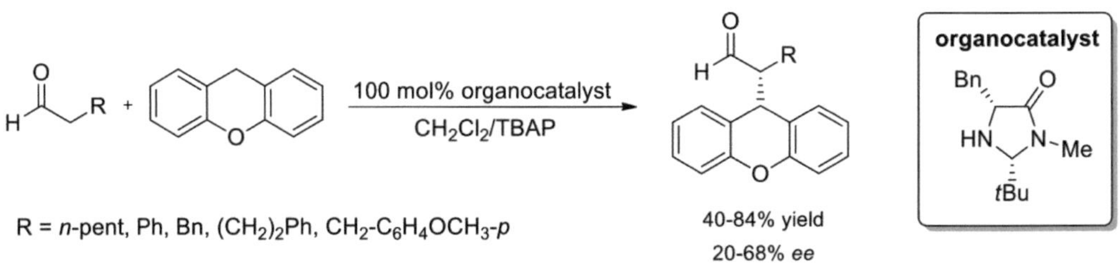

Scheme 15.

An asymmetric electro-organocatalysis method for enantioselective α-alkylation of aldehydes with xanthene has also been devised by the group of Jang et al. [69]. **Scheme 15** depicts the best results using a chiral imidazole as organocatalyst, which was chosen from a plethora of differently substituted imidazole-based compounds [69]. According to electrochemical studies and control experiments, the reaction is probable to occur through the formation of an enamine intermediate. DFT calculations suggested that xanthene adds to the opposite side of the phenyl ring of the radical intermediate blue to stereochemical hindrance issues, thus enhancing the stereoselectivity of the reaction.

In 2014, Xu and co-workers [70] published an electrochemical intramolecular aminooxygenation reaction of unactivated alkenes based on the addition of N-centered radicals to alkenes (generated from electrochemical oxidation) followed by trapping of the cyclized radical intermediate with TEMPO. This process allowed the preparation of different aminooxygenation products in high yields and excellent trans-selectivity for cyclic systems (d.r. up to > 20:1).

Very recent, Xu and collaborators [71] reported the first electrocatalytic method using ferrocene as a cheap redox catalyst to produce amidyl radicals from N-arylamides. The conventional methods for oxidative generation of amidyl radicals from N—H amides need to use a stoichiometric quantity of expensive noble-metal catalysts or strong oxidants. In this case, the authors showed an efficient radical-generating process based on intramolecular olefin hydroamidation reaction.

Acknowledgements

This work was supported by Fundação para a Ciência e a Tecnologia through projects (PEst-C/LA0006/2013) one contract under principal investigator FCT (L.C. Branco) and one post-doctoral fellowships (Hugo Cruz—SFRH/BPD/102705/2014).

Author details

Karolina Zalewska, Miguel M. Santos, Hugo Cruz and Luis C. Branco*

*Address all correspondence to: l.branco@fct.unl.pt

Department of Chemistry, Faculty of Science and Technology, LAQV-REQUIMTE, New University of Lisbon, Caparica, Portugal

References

[1] Belle M, Bolm C, editors. Transition Metals for Organic Synthesis. 2nd ed. Weinheim: Wiley-VCH; 2004. doi: 10.1002/9783527619405

[2] Wessig P. Organocatalytic enantioselective photoreaction. Angew. Chem. Int. Ed. 2006; 45:2168–2171. doi: 10.1002/anie.200503908

[3] Ravelli D, Fagnoni M, Albini A. Photoorganocatalysis. What for?. Chem. Soc. Rev. 2013; 42:97–113. doi: 10.1039/C2CS35250H

[4] Wang C, Lu Z. Catalytic enantioselective organic transformations via visible light photocatalysis. Org. Chem. Front. 2015; 2:179–190. doi: 10.1039/C4QO00306C

[5] Fagnoni M, Dondi D, Ravelli D, Albini A. Photocatalysis for the formation of the C—C bond. Chem. Rev. 2007; 107:2725–2756. doi: 10.1021/cr068352x

[6] Turro NJ, Ramamurthy V, Scaiano JC. Modern Molecular Photochemistry of Organic Molecules. USA: University Science Books; 2010.

[7] Arceo E, Jurberg ID, Alvarez-Fernandez A, Melchiorre P. Photochemical activity of a key donor-acceptor complex can drive stereoselective catalytic-alkylation of aldehydes. Nat. Chem. 2013; 5:750–756. doi: 10.1038/nchem.1727

[8] Nicewicz DA, MacMillan DWC. Merging photoredox catalysis with organocatalysis: the direct asymmetric alkylation of aldehydes. Science. 2008; 322:77–80. doi: 10.1126/science.1161976

[9] Wiegand C, Herdtweck E, Bach T. Enantioselectivity in visible light-induced, singlet oxygen [2+4] cycloaddition reactions (type II photooxygenations) of 2-pyridones. Chem. Commun. 2012; 48:10195–10197. doi: 10.1039/C2CC35621J

[10] Bauer A, Westkämper F, Grimme S, Bach T. Catalytic enantioselective reactions driven by photoinduced electron transfer. Nature. 2005; 436:1139–1140. doi: 10.1038/nature03955

[11] Jeon YT, Lee CP, Mariano PS. Radical cyclization reactions of silyl amine α,β-unsaturated ketone and ester systems promoted by single electron transfer photosensitization. J. Am. Chem. Soc. 1991; 113:8847–8863. doi: 10.1021/ja00023a038

[12] Müller C, Bauer A, Bach T. Light-driven enantioselective organocatalysis. Angew. Chem. Int. Ed. 2009; 48:6640–6642. doi: 10.1002/anie.200901603

[13] Müller C, Bauer A, Maturi MM, Cuquerella MC, Miranda MA, Bach T. Enantioselective intramolecular [2+2]-photocycloaddition reactions of 4-substituted quinolones catalyzed by a chiral sensitizer with a hydrogen-bonding motif. J. Am. Chem. Soc. 2011; 133:16689–16697. doi: 10.1021/ja207480q

[14] Breitenlechner S, Bach T. A polymer-bound chiral template for enantioselective photochemical reactions. Angew. Chem. Int. Ed. 2008; 47:7957–7959. doi: 10.1002/anie.200802479

[15] Guo H, Herdtweck E, Bach T. Enantioselective lewis acid catalysis in intramolecular [2+2] photocycloaddition reactions of coumarins. Angew. Chem. Int. Ed. 2010; 49:7782–7785. doi: 10.1002/anie.201003619

[16] Vallavoju N, Selvakumar S, Jockusch S, Sibi MP, Sivaguru J. Enantioselective organophotocatalysis mediated by atropisomeric thiourea derivatives. Angew. Chem. Int. Ed. 2014; 53:5604–5608. doi: 10.1002/anie.201310940

[17] Jensen KL, Dickmeiss G, Jiang H, Albrecht Ł, Jørgensen KA. The diarylprolinol silyl ether system: a general organocatalyst. Acc. Chem. Res. 2012; 45:248–264. doi: 10.1021/ar200149w

[18] Arceo E, Bahamonde A, Bergonzinib G, Melchiorre P. Enantioselective direct α-alkylation of cyclic ketones by means of photo-organocatalysis. Chem. Sci. 2014; 5:2438–2442. doi: 10.1039/C4SC00315B

[19] Silvi M, Arceo E, Jurberg ID, Cassani C, Melchiorre P. Enantioselective organocatalytic alkylation of aldehydes and enals driven by the direct photoexcitation of enamines. J. Am. Chem. Soc. 2015; 137:6120–6123. doi: 10.1021/jacs.5b01662

[20] Barata-Vallejo S, Flesia MM, Lantaño B, Argüello JE, Peñéñory AB, Postigo A. Heterogeneous photoinduced homolytic aromatic substitution of electron-rich arenes with perfluoroalkyl groups in water and aqueous media – a radical ion reaction. Eur. J. Org. Chem. 2013; 5:998–1008. doi: 10.1002/ejoc.201201271

[21] Woźniak Ł, Murphy JJ, Melchiorre P. Photo-organocatalytic enantioselective perfluoroalkylation of β-ketoesters. J. Am. Chem. Soc. 2015; 137:5678–5681. doi: 10.1021/jacs.5b03243

[22] Lian M, Li Z, Cai Y, Meng Q, Gao Z. Enantioselective photooxygenation of β-keto esters by chiral phase-transfer catalysis using molecular oxygen. Chem. Asian J. 2012; 7:2019–2023. doi: 10.1002/asia.201200358

[23] Schuster DI. The photochemistry of enones. In: Patai S, Rappoport Z, editors. The Chemistry of Enones. New York: Wiley. 1989. p. 623–756

[24] Chanon M, Eberson L, Conceptual basis. In: Fox MA, Chanon M, editors. Photoinduced Electron Transfer. New York: Elsevier. 1988.

[25] Arceo E, Montroni E, Melchiorre P. Photo-organocatalysis of atom-transfer radical additions to alkenes. Angew. Chem. Int. Ed. 2014; 53:12064–12068. doi: 10.1002/anie.201406450

[26] Ravelli D, Fagnoni M. Aromatic aldehydes as energy-transfer photoorganocatalysts. Chem. Cat. Chem. 2015; 7:735–737. doi: 10.1002/cctc.201403024

[27] Kandukuri SR, Bahamonde A, Chatterjee I, Jurberg ID, Escudero EC, Melchiorre P. X-ray characterization of an electron donor-acceptor complex that drives the photochemical alkylation of indoles. Angew. Chem. Int. Ed. 2015; 54:1485–1489. doi: 10.1002/anie.201409529

[28] Neumann M, Füldner S, König B, Zeitler K. Metal-free, cooperative asymmetric organophotoredox catalysis with visible light. Angew. Chem. Int. Ed. 2011; 50:951–954. doi: 10.1002/anie.201002992

[29] Neumann M, Zeitler K. Application of microflow conditions to visible light photoredox catalysis. Org. Lett. 2012; 14:2658–2661. doi: 10.1021/ol3005529

[30] Fidaly K, Ceballos C, Falguières A, Veitia MSI, Guy A, Ferroud C. Visible light photoredox organocatalysis: a fully transition metal-free direct asymmetric α-alkylation of aldehydes. Green Chem. 2012; 14:1293–1297. doi: 10.1039/C2GC35118H

[31] Cecere G, König CM, Alleva JL, MacMillan DWC. Enantioselective direct α-amination of aldehydes via a photoredox mechanism: a strategy for asymmetric amine fragment coupling. J. Am. Chem. Soc. 2013; 135:11521–11524. doi: 10.1021/ja406181e

[32] Cassani C, Bergonzini G, Wallentin CJ. Photocatalytic decarboxylative reduction of carboxylic acids and its application in asymmetric synthesis. Org. Lett. 2014; 16:4228–4231. doi: 10.1021/ol5019294

[33] Yang Z, Li H, Zhang L, Zhang MT, Cheng JP, Luo S. Organic photocatalytic cyclization of polyenes: a visible-light-mediated radical cascade approach. Chem. Eur. J. 2015; 21:14723–14727. doi: 10.1002/chem.201503118

[34] Goetz AE, Boydston AJ. Metal-free preparation of linear and cross-linked polydicyclopentadiene. J. Am. Chem. Soc. 2015; 137:7572–7575. doi: 10.1021/jacs.5b03665

[35] Fukuzumi S, Ohkubo K. Organic synthetic transformations using organic dyes as photoredox catalysts. Org. Biomol. Chem. 2014; 12:6059–6069. doi: 10.1039/C4OB00843J

[36] Weber M, Weber M, Kleine-Boymann M. Phenol. Ullmann's Encyclopedia of Industrial Chemistry, Weinheim: Wiley-VCH, 2004. doi: 10.1002/14356007.a19_299.pub2

[37] Ohkubo K, Kobayashi T, Fukuzumi S. Photocatalytic alkoxylation of benzene with 3-cyano-1-methylquinolinium ion. Opt. Express. 2012; 20:A360–A365. doi: 10.1364/OE.20.00A360

[38] Furuya T, Adam S, Kamlet AS, Ritter T. Catalysis for fluorination and trifluoromethylation. Nature. 2011; 473:470–477. doi: 10.1038/nature10108

[39] Ohkubo K, Fujimoto A, Fukuzumi S. Visible-light-induced oxygenation of benzene by the triplet excited state of 2,3-dichloro-5,6-dicyano-p-benzoquinone. J. Am. Chem. Soc. 2013; 135:5368–5371. doi: 10.1021/ja402303k

[40] Fukuzumi S, Kotani H, Ohkubo K, Ogo S, Tkachenko NV, Lemmetyinen H. Electron-transfer state of 9-mesityl-10-methylacridinium ion with a much longer lifetime and higher energy than that of the natural photosynthetic reaction center. J. Am. Chem. Soc. 2004; 126:1600–1601. doi: 10.1021/ja038656q

[41] Ohkubo K, Nanjo T, Fukuzumi S. Efficient photocatalytic oxygenation of aromatic alkene to 1,2-dioxetane with oxygen via electron transfer. Org. Lett. 2005; 7:4265–4268. doi: 10.1021/ol051696+

[42] Hari DP, Konig B. Synthetic applications of eosin Y in photoredox catalysis. Chem. Commun. 2014; 50:6688–6699. doi: 10.1039/C4CC00751D

[43] Hedstrand DM, Kruizinga WH, Kellogg RM. Light induced and dye accelerated reductions of phenacyl onium salts by 1,4-dihydropyridines. Tetrahedron Lett. 1978; 19:1255–1258. doi: 10.1016/S0040-4039(01)94515-0

[44] Cantillo D, de Frutos O, Rincon JA, Mateos C, Kappe CO. Continuous flow α-trifluoromethylation of ketones by metal-free visible light photoredox catalysis. Org. Lett. 2014; 16:896–899. doi: 10.1021/ol403650y

[45] Yang XJ, Chen B, Zheng LQ, Wu LZ, Tung CH. Highly efficient and selective photocatalytic hydrogenation of functionalized nitrobenzenes. Green Chem. 2014; 16:1082–1086. doi: 10.1039/C3GC42042F

[46] Yang DT, Meng QY, Zhong JJ, Xiang M, Liu Q, Wu LZ. Metal-free desulfonylation reaction through visible-light photoredox catalysis. Eur. J. Org. Chem. 2013; 33:7528–7532. doi: 10.1002/ejoc.201301105

[47] Hari DP, Konig B. Eosin Y catalyzed visible light oxidative C—C and C—P bond formation. Org. Lett. 2011; 13:3852–3855. doi: 10.1021/ol201376v

[48] Liu Q, Li YN, Zhang HH, Chen B, Tung CH, Wu LZ. Reactivity and mechanistic insight into visible-light-induced aerobic cross-dehydrogenative coupling reaction by organophotocatalysts. Chem. Eur. J. 2012; 18:620–627. doi: 10.1002/chem.201102299

[49] Hari DP, Schroll P, König B. Metal-free, visible-light-mediated direct C—H arylation of heteroarenes with aryl diazonium salts. J. Am. Chem. Soc. 2012; 134:2958–2961. doi: 10.1021/ja212099r

[50] Savéant JM. Molecular catalysis of electrochemical reactions. Mechanistic aspects. Chem. Rev. 2008; 108:2348–2378. doi: 10.1021/cr068079z

[51] Yoshida J, Kataoka K, Horcajada R, Nagaki A. Modern strategies in electroorganic synthesis. Chem. Rev. 2008; 108:2265–2299. doi: 10.1021/cr0680843

[52] Gallardo I, Guirado G. Thermodynamic study of σ^H-complexes in nucleophilic aromatic substitution reactions: relative stabilities of electrochemically generated radicals. Eur. J. Org. Chem. 2008; 14:2463–2472. doi: 10.1002/ejoc.200701098

[53] Moeller KD. Synthetic applications of anodic electrochemistry. Tetrahedron. 2000; 56:9527–9554. doi: 10.1016/S0040-4020(00)00840-1

[54] Sperry JB, Wright DL. The application of cathodic reductions and anodic oxidations in the synthesis of complex molecules. Chem. Soc. Rev. 2006; 35:605–621. doi: 10.1039/B512308A

[55] Heinze J. Cathodic reactions of Hydrocarbons in Organic Electrochemistry. Ed. H. Lund, O. Hammerich. Marcel Dekker: New York. 2001.

[56] Francke R, Little RD. Redox catalysis in organic electrosynthesis: basic principles and recent developments. Chem. Soc. Rev. 2014; 43:2492–2521. doi: 10.1039/C3CS60464K

[57] Gallardo I. Nucleophilic aromatic substitution of hydrogen: a novel electrochemical approach to the cyanation of nitroarenes. Chem. Eur. J. 2001; 7:1759–1765 doi: 10.1002/1521-3765(20010417)7:8<1759::AID-CHEM17590>3.0.CO;2-F

[58] Cruz H, Gallardo I, Guirado G. Electrochemical synthesis of organophosphorus compounds through nucleophilic aromatic substitution: mechanistic investigations and synthetic scope. Eur. J. Org. Chem. 2011; 36:7378–7389. doi: 10.1002/ejoc.201101357

[59] Gallardo I, Guirado G, Marquet J. Nucleophilic aromatic substitution for heteroatoms: an oxidative electrochemical approach. J. Org. Chem. 2002; 67:2548–2555. doi: 10.1021/jo010847t

[60] Alfonso-Súarez P, Kolliopoulos VA, Smith JP, Banks CE, Jones AM. An experimentalist's guide to electrosynthesis: the Shono oxidation. Tetrahedron Lett. 2015; 56:6863–6867. doi: 10.1016/j.tetlet.2015.10.090

[61] Cruz H, Gallardo I, Guirado G. Electrochemically promoted nucleophilic aromatic substitution in room temperature ionic liquids—an environmentally benign way to functionalize nitroaromatic compounds. Green Chem. 2011; 13:2531–2542. doi: 10.1039/C1GC15303J

[62] Ogawa KA, Boydston AJ. Recent developments in organocatalyzed electroorganic chemistry. Chem. Lett. 2015; 44:10–16. doi: 10.1246/cl.140915

[63] Zeng CC, Zhang NT, Lam CM, Little RD. Novel triarylimidazole redox catalysts: synthesis, electrochemical properties, and applicability to electrooxidative C—H activation. Org. Lett. 2012; 14:1314–1317. doi: 10.1021/ol300195c

[64] Gao WJ, Li WC, Zeng CC, Tian HY, Hu LM, Little RD. Electrochemically initiated oxidative amination of benzoxazoles using tetraalkylammonium halides as redox catalysts. J. Org. Chem. 2014; 79:9613–9618. doi: 10.1021/jo501736w

[65] Chen J, Yan WQ, Lam CM, Zeng CC, Hu LM, Little RD. Electrocatalytic aziridination of alkenes mediated by n-Bu_4NI: a radical pathway. Org. Lett. 2015; 17:986–989. doi: 10.1021/acs.orglett.5b00083

[66] Jensen KL, Franke PT, Nielsen LT, Daasbjerg K, Jørgensen KA. Anodic oxidation and organocatalysis: direct regio- and stereoselective access to meta-substituted anilines by α-arylation of aldehydes. Angew. Chem. Int. Ed. 2010; 49:129–133. doi: 10.1002/anie.200904754

[67] Schämann M, Schäfer HJ. Reaction of enamines and mediated anodic oxidation of carbohydrates with the 2,2,6,6-tetramethylpiperidine-1-oxoammonium ion (TEMPO$^+$). Electrochim. Acta. 2005; 50:4956–4972. doi: 10.1016/j.electacta.2005.02.077

[68] Bui NN, Ho XH, Mho S, Jang HY. Organocatalyzed α-oxyamination of aldehydes using anodic oxidation. Eur. J. Org. Chem. 2009; 31:5309–5312. doi: 10.1002/ejoc.200900871

[69] Ho XH, Mho S, Kang H, Jang HY. Electro-organocatalysis: enantioselective α-alkylation of aldehydes. Eur. J. Org. Chem. 2010; 23:4436–4441. doi: 10.1002/ejoc.201000453

[70] Xu F, Zhu L, Zhu S, Yan X, Xu HC. Electrochemical intramolecular aminooxygenation of unactivated alkenes. Chem. Eur. J. 2014; 20:12740–12744. doi: 10.1002/chem.201404078

[71] Zhu L, Xiong P, Mao ZY, Wang YH, Yan X, Lu X, Xu HC. Electrocatalytic generation of amidyl radicals for olefin hydroamidation: use of solvent effects to enable anilide oxidation. Angew. Chem. Int. Ed. 2016; 55:2226–2229. doi: 10.1002/anie.201510418

Section 4

Recent Advances in Organocatalysis

Chapter 7

Recent Advances in Sustainable Organocatalysis

Luis C. Branco, Ana M. Faisca Phillips,
Maria M. Marques, Sandra Gago and
Paula S. Branco

Additional information is available at the end of the chapter

http://dx.doi.org/10.5772/64279

Abstract

The recent advances on green and sustainable organocatalysis are revised in this chapter. An important focus on one of the 12 principles of green chemistry, organocatalysis pursues to reduce energy consumption as well as to optimize the use of different resources, targeting to become a sustainable strategy in organic chemical transformations. In last decades, several experimental methodologies have been performed to make organocatalysis an even greener and sustainable alternative to stoichiometric approaches as well as non-catalytic conditions by the use of benign and friendlier reaction media. In this line, several approaches using water as preferential solvent, alternative solvents such as ionic liquids including chiral ones, deep eutectic solvents, polyethylene glycol (PEG), supercritical fluids and organic carbonates or solvent-free methodologies have been reported. In this chapter, we mainly focus on the recent remarkable advancements in organocatalysis using green and sustainable protocols.

Keywords: water, solvent-free, heterogeneous organocatalysis, alternative solvents, sustainable organocatalysis

1. Introduction

Aqueous reactions can combine the unique physical properties exhibited by water and other desirable advantages from the point of view of environmental concerns, safety and low cost. For many cases, the application of water is limited due to the reduced solubility and stability of diverse organic substrates in aqueous media as well as the problems associated with possible contamination of water phase with organic substrates and the need of efficient purification steps.

Alternatively, the strategy to use solvent-free organocatalysed reactions has also been explored. In these cases, a large excess of reagent, which acts as reaction media avoiding the additional use of auxiliary solvents is tested. Alternative solvents such as ionic liquids (ILs); acyclic and cyclic carbonate and polycarbonate solvents (e.g. dimethylcarbonate and polyethylenoglycol derivatives); fluorinated solvents and supercritical carbon dioxide have been largely explored for catalytic processes. In this context, the large number of examples is centred in the application of ionic liquids including chiral ones as efficient reaction media for several organic transformations. Taking advantage of physical, chemical and thermal properties of ionic liquids as well as the possibility to tune their properties according to the adequate cation-anion combinations, it is possible to develop designer solvents for organocatalytic reactions. Also, the high capacities to solubilize and stabilize different organic, inorganic and polymeric materials as well as the reusable and recyclable behaviour are relevant parameters to justify the large application of ionic liquids in synthesis and catalysis. Recent advances showed the potential use of supercritical carbon dioxide as unique solvent or in combination with ionic liquids for efficient reaction-extraction processes. Supported catalytic processes as efficient, greener and recyclable methodologies for some organocatalytic processes have been also described. The organocatalysts including chiral ones can be incorporated in solid supports improving their stability and catalytic activity as well as the possibility to reuse and recycle several times without significant decrease in their performance. The use of alternative techniques such as microwave irradiation and ultra-sons (sonication) instead of the traditional synthetic protocols will be also reviewed. One of the most important challenges of synthetic chemistry is related with the combination of efficiency, reduced costs and environmental impact in the production of relevant molecules, particularly for the preparation of chiral compounds. The creation of chiral centres can be achieved by several methodologies such as by using chiral auxiliaries, readily obtained by chemical manipulation of chiral natural and non-natural compounds by asymmetric catalysis including biocatalysis.

2. Organocatalytic reactions in water

Traditionally, the majority of organic reactions have been performed in organic solvents, mainly due to the fact that most organic compounds are not very water-soluble. In addition, many reagents used in organic synthesis are destroyed by water. This fact is contradictory to what happens in nature, where reactions promoted by enzymes and antibodies take place in aqueous media. Despite their utility in solubilising substrates and reagents, organic solvents are toxic and volatile. Water has a few obvious advantages over organic solvents [1–4]; it is relatively abundant, non-toxic, non-flammable and inexpensive. In addition, it has a large temperature window in which it remains in the liquid state and high heat capacity, making it a good and safe heat sink for exothermic reactions, particularly important when they are carried out on a large scale. Water also has a large dielectric constant, high surface tension, hydrogen bonding capacity and optimum oxygen solubility. Since the solvent is usually present in large excess, it can play an important role in the reaction. Initially, it was thought that the presence of water was only detrimental to organic reactions [5]; it brought insolubility

http://dx.doi.org/10.5772/64279

problems, and it could react with functional groups from different substrates, slowing them down and causing low yields. Due to its capacity to form hydrogen bonds, water could disrupt H-bonding in transition states, i.e. those formed between catalyst and substrate molecules, deteriorating catalytic activity and stereocontrol. Pioneering studies by Breslow showed that Diels-Alder reactions were accelerated in water [6]. This fact was a surprise, since Diels-Alder reactions are relatively insensitive to solvent polarity. He determined that the acceleration was due to the fact that in the presence of water, the less polar reagents would be drawn together in hydrophobic hydration, resulting in a more favourable overall entropy. The increase in concentration led to rate enhancements. In 2005, Sharpless showed that many uni- and bimolecular reactions were accelerated when carried out in vigorously stirred aqueous media [7]. Reactions which took place as an emulsion displaying rate acceleration, he described as taking place 'on water'. It would later be observed that under the right conditions the benefits of water may be not only to rate acceleration but also to increased selectivity (vide infra). The effect of water on organocatalytic reactions was investigated even when the first examples of this type of catalysis were studied, i.e. in the direct proline-catalysed aldol reaction (**Scheme 1**). This reaction, in which two unmodified carbonyl compounds react to give a β-hydroxy

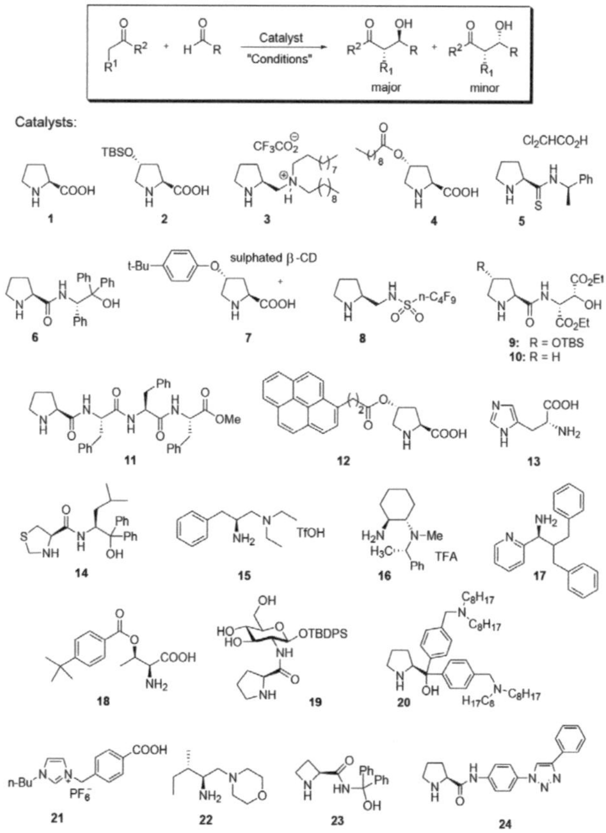

Scheme 1. The direct aldol reaction and a selection of catalysts which perform well in aqueous media.

carbonyl product, is used by aldolase enzymes for the biosynthesis of carbohydrates, keto acids and some amino acids. In organic synthesis, the organocatalytic direct aldol version is catalysed by chiral amines via *in situ* formation of imines and then enamines, in a mechanism similar to that used by aldolase Type I enzymes and antibodies. This C-C bond-forming reaction is extremely useful and it may be used for the stereoselective assembly of complex polyols. Nowadays, the use of water as a solvent for few asymmetric organocatalytic reactions was already described, but studies on the aldol reaction predominate by far in the literature [8–10].

2.1. The direct aldol reaction

The capability of proline (1) to promote asymmetric direct intra-molecular aldol reactions was shown in the Hajos-Parrish-Eder-Sauer-Wiechert cyclisation in 1971. However, despite the utility of this reaction for the synthesis of steroids, it would be only 30 years later that the broad applicability of this catalyst would be discovered and the first inter-molecular version of the direct aldol was described by List and Barbas [11]. High yields and stereoselectivities can be obtained with proline catalysis in organic solvents, but addition of water to the reaction mixture lowers the yields and stereoselectivities, not only in these examples but also in other studies reported in subsequent years. In the asymmetric direct aldol reaction, the chemo-, regio-, diastereo- and enantioselectivity should be controlled. Side reactions are possible, which reduce the yield: aldol condensation, aldol reaction and condensation of the aldehyde acceptor and also oxazolidinone formation between the catalyst and the aldehyde. Excess ketone is often used to prevent aldehyde homodimerisation and catalyst kill events. Another inherent problem is that aldol reactions are very difficult to achieve under stoichiometric conditions, since the equilibrium constants for many direct ketone-aldehyde aldol reactions are just barely on the side of the products. Pihko and co-workers were the first to observe significant rate enhancements, yield and stereoselectivity increases in the proline-catalysed aldol reaction when small amounts of water (1–10 equiv.) were added to a DMF solution of acetone or 4-thianone and various aromatic aldehydes [12]. The effect obtained was such that the reaction could be performed with equimolar quantities of reagents. The beneficial role of water was attributed to a suppression of the competing reaction which leads to oxazolidinone formation. The first direct aldol reactions performed solely in water were later described independently by Hayashi [13] and Barbas [14] in 2006. To achieve their aim, these authors developed novel organocatalysts derived from proline containing hydrophobic groups. In these reactions, aldehyde and ketone substrate molecules aggregate excluding water, thus generating a two-phase system. Hayashi's catalyst, 4-siloxyproline 2 (**Scheme 1**) operates in the organic phase where enamine formation takes place. A reaction in this heterogeneous system was defined by Hayashi as a '*direct aldol reaction in the presence of water*' as opposed to a system in which the reactants are dissolved in water, which he labelled '*reactions in water*'. In these reactions, water is not an inert second phase, but it influences the stereoselectivity and yield, although an explanation for the effect obtained was not presented. Barbas and co-workers performed the direct cross-aldol reaction of cyclohexanone (2 equiv.) with 4-nitro-benzaldehyde in pure water, obtaining the *anti*-aldol product in 94% *ee*. Catalyst loading could be as low as 1 mol% with protonated diamine 3. This catalyst is a water soluble quaternary ammonium salt with long hydrophobic chains, and it probably worked at the water-organic interface of the emulsion formed, in a similar way than organic surfactants.

In an emulsion, hydrophobic catalysts also reduce the contacts between the transition state and bulky water. Catalysts 2 and 3 were less efficient in reactions with non-activated aldehydes showing lower yields and *ees* of products. Hayashi also developed catalyst 4 for cross-aldol reactions of aldehydes in water [15]. In the presence of 5 equiv. of aliphatic aldehyde, products were isolated in moderate to high yields, high *drs* and *ees* (89–99%).

The concepts applied in the development of catalysts 2–4 would also be used for catalyst design in subsequent years (**Scheme 1**). Proline analogues obtained through derivatisation of the carboxyl or the amino groups and 4-hydroxyproline derivatives have predominated. These reactions are usually *anti*-selective while the use of acyclic amino acids and analogues usually promote *syn*-selective reactions.

Due to space limitations, in the examples the highlighted solvent was usually pure water without the addition of any organic co-solvents, although in some cases one of the carbonyl components was used in excess and could play the role of a solvent too.

Gryko studied in more detail hydrophobic aggregation processes [16]. The use of different salting-in and salting-out conditions in the reaction between cyclic ketones and aromatic aldehydes catalysed by protonated thioamide 5 showed that both the reaction rate and stereochemistry were affected by the rate of hydrophobic aggregation. Around this time, Singh described a very efficient prolinamide (6) for *anti*-aldol reactions between ketones and aldehydes in brine [17]. As little as 0.5 mol% provided very high *ees* even in reactions of acetone. The efficiency of this catalyst was ascribed to its capability to activate the acceptor aldehyde through the formation of hydrogen bonds via the amine and hydroxyl groups and also to a salting-out effect. Armstrong emphasized product recovery and developed protected hydroxyproline 7 which, when used in water with a sulphated β-cyclodextrin, allowed the formation of *anti*-aldol products from equimolar amounts of cyclohexanone and aryl aldehydes. The products could be obtained in quantitative yields after a simple filtration [18]. Catalyst recovery and recycling was possible with Wang's fluorous proline sulphonamide 8, developed for aldol reactions between ketones or aldehydes with aromatic aldehydes on water. The catalyst was recovered by fluorous extraction and reused up to seven cycles [19].

In 2008, Gong developed highly reactive prolinamide 9, showing that only 1 mol% was enough to catalyse the direct aldol between a wide range of aromatic aldehydes and 2 equiv. of cyclic or linear ketones in water to afford products in high yields and *ees* (91–99%) [20]. For direct aldol reactions of hydroxyacetone and fluoroacetone with electron-poor aromatic aldehydes, the related prolinamide 10 (20–30 mol%) was developed [21]. However, for hydroxyacetone products (with 84–96% *ees*), THF/water mixtures (2:1) were required. Chiral 1,4-diols, which are disfavoured products in aldol reactions catalysed by aldolases or L-proline, were obtained as the major products. This regioselectivity was possible only when water is presented. Similarly, fluoroacetone would only react when water was added to the THF solution. Theoretical studies revealed that water forms hydrogen bonds with the amide oxygen of the prolinamide and the hydroxyl of hydroxyacetone, which influence the regioselectivity by micro-solvation. Previously, the same group had used small proline-based peptides, e.g. catalyst 11, which also worked well provided water was presented [22]. Contrary to what is observed with cyclohexanone, the aldol reactions of cyclopentanone are more difficult, and often the *drs* obtained are low. As little as 2 mol% of catalyst 12, developed by Gruttadauria in 2008, catalysed reactions with aromatic aldehydes, giving *ees* in the range 93–99% [23]. The solvent was pure water, but 5 equiv. of ketone were used. In the same year, Zhao achieved the direct cross-aldol reaction between ketones and β, γ-unsaturated ketoesters with 2 [24]. Two chiral centres, one quaternary, were assembled in 98 to more than 99% *ee* when cyclohexanone was the selected substrate. The reaction was faster and the yield and stereoselectivities were higher in water. Other ketones provided lower *ees*, but *dr* remained very high (≥19:1). Cross-reactions of aldehydes are particularly problematic to achieve because of the formation of various side products, in particular the self-aldolisation is very difficult to control. When secondary amines are used, anti-products are obtained. The first cross-aldol reaction in which α-branched aldehydes were reacted as the ene component was reported by Mahrwald in 2009 [25]. Under D-histidine (13) catalysis, *syn* β-hydroxyaldehydes were obtained in high yields and very high *ees*. Electron-rich aldehydes reacted in water exclusively as the ene component and electron-poor aldehydes as the carbonyl component, allowing for the formation of quaternary carbon centres. Aldol methodologies were applied for the first time to total synthesis of pantolactone and lyxose. Also in 2009, Singh developed highly reactive L-cysteine-derived 14, which promoted even aldol reactions of acetone (a problematic substrate) in brine, in quantities as low as 1 mol%, with several aromatic aldehydes affording 70–86% yields, 96–99% *ees* [26].

Despite all these developments, during the next five years the organocatalytic direct aldol reaction still continued to attract considerable attention and catalysts with novel backbones were reported. Luo used primary-tertiary diamine Brønsted acid 15 to obtain the otherwise difficult to get *syn* aldol products, isolated after borohydride reduction [27]. Good results were reported with linear and branched aldehydes. Glycoaldehyde donors were also used, but in DMF, although it was observed for the first time for these substrates that addition of water caused large improvements in *ee*.

In an early study on the cross-aldol reaction of ketones catalysed by proline, Garden and co-workers explored the effect of water [28]. They found that when excess acetone was reacted neat with isatins, which may be viewed as a type of activated ketones, addition of small amounts of water gave large enhancements in yields and *ees*. In 2010, Singh showed that cross-aldol reaction between cyclohexanone and five substituted isatins is possible in water in the presence of primary-tertiary diamine Brønsted acid 16 [29]. The *syn* products, which are potential anti-convulsants, were obtained in excellent *drs* and high *ees* in either DMF or water. The majority of catalysts tried during this period were based on amino acids. Nugent used a different template to obtain 2-pycolylamine 17 [30]. It performed very well affording 89–99% *ees* in reactions between cyclic ketones, including *N*-Boc-piperidone and aromatic aldehydes, in water or brine, in the presence of 2,4-dinitrobenzenesulphonic acid. Simple O-acylation of threonine provided novel surfactant organocatalysts, e.g. 18 [31]. When tested by Fu on reactions between cyclic or aliphatic ketones and aromatic aldehydes, high yields and stereoselectivities were described. Even unprotected hydroxyacetone could be reacted in water with nitrobenzaldehydes to provide very good results too. The catalyst could be recovered and reused without loss of performance when tried on a large-scale reaction (25 mmol aldehyde). In the same year, Caputo showed that D-glucosamine could be used successfully as a catalyst template when coupled with L-proline, e.g. in 19, to promote the direct aldol reaction between cyclohexanone and substituted benzaldehydes in brine [32]. Catalytic loading could be as low as 2 mol% to provide very high *ees* (82–99%). Ni, Headley and co-workers were the first to achieve successfully the direct asymmetric cross-aldol reaction of acetaldehyde (10 equiv.), a difficult substrate prone to undergo side reactions with various aromatic aldehydes in aqueous media [33]. To catalyse the reaction, a diaryl prolinol containing dioctylamino groups (20) (5 mol%) was used, with ionic liquid 21 (10 mol%) as co-catalyst. The products were isolated in high yields and *ees* after borohydride reduction. Also, in 2013 the catalyst loading in cross-aldol reactions of ketones in water was lowered to a new minimum value: from 5 to 1 mol%. This was possible with new primary-tertiary diamine catalyst 22 developed by Chimni for reactions of isatins [34]. In 2014, Wang developed azetidine-2-carboxamides 23 as catalysts [35], which were tested on aldol reactions between acetone and benzaldehydes in brine. The yields varied in the range of 38–88%, being lower with more demanding substrates (67–96% *ees*). In the same year, Dash developed a novel approach to obtain tubuvaline precursors based on a prolinamide 24 catalysed aldol reaction of thiazole carbaldehydes with ketones 'on water' [36]. Tubuvalin is the core structure of a family of tetrapeptides, the tubulysins, which are the most potent anti-cancer agents known so far. The best results obtained with methyl isopropyl ketones donors were when 20 mol% catalyst was used together with two acid additives, formic acid and 2,4,6-trifluorobenzoic acid. The formation of a branched product competed, but when the reaction was performed in water, the linear aldol product was obtained with >99:1 regioselectivity and 92% *ee*. The authors reasoned that the facial linear selectivity was due to the preferential formation of the least substituted enamine, since the more substituted one is thermodynamically less favourable due to greater steric congestion. In this system, the reactants and the catalyst remain undissolved during the reaction. When a soluble ketone, acetone, was reacted, the product was racemic, but an enantioselective reaction was possible when the catalyst was changed to proline with *R*-BINOL as co-catalyst.

2.2. The Mannich reaction

The asymmetric Mannich reaction is one of the most important methods for the synthesis of enantioenriched molecules containing a stereogenic carbon-nitrogen bond. Of lately, there have been several reports on Mannich reactions performed in organic solvents to which small amounts of water were added [1, 37]. Reported methods using water as unique reaction media are few, probably because this is more difficult to achieve due to the inherent susceptibility of the imine substrates to hydrolysis [1]. Enamine catalysis was used in all cases. The first reaction performed solely in water was described by Ibrahim and Córdova in 2006 [38]. They showed that 10 mol% of TMS-protected prolinol **25** catalysed the Mannich reaction between aldehydes (3 equiv.) and glyoxylate imines, to afford *anti*-products in moderate yields but very high stereoselectivities (97–99% *ee*).

In 2007, a three-component Mannich reaction of O-benzyl hydroxyacetone, *p*-anisidine and aliphatic or aromatic aldehydes, taking place solely in water, was described by Lu [39]. The reaction was promoted by primary amino acid, derived from L-threonine, and also led to the formation of *O*-protected *anti*-1,2-amino alcohols in good to excellent yields and high *ees*. Amedjkouh and Brandberg described an interesting example of autocatalysis based on a direct Mannich reaction when cyclohexanone was reacted with *N*-PMP-protected glyoxylate imine [40]. The product was found to act as a catalyst for its own replication and the *ees* obtained in aqueous solutions (pH 7) were much better than those obtained in organic solvents. The major product was *syn*-configured. Teo, Lau and Wu found that 5–10 mol% of a TBDPS-protected L-serine could catalyse the three-component Mannich reaction between ketones, aldehydes and *p*-anisidine in water. In this example, *syn*-products were obtained with aliphatic aldehydes, but *anti*-products were produced from cyclic ketones [41]. Protected hydroxyl proline was found by Hayashi to promote *syn*-selective Mannich reactions between linear aldehydes and glyoxylate imines. The reactions were more favourable in basic solutions made with sodium bicarbonate and the products could be obtained in good yields and high stereoselectivities [42]. For the three-component reaction of dimethoxyacetaldehyde with cyclohexanone and *p*-anisidine, siloxy-tetrazole hybrid catalyst afforded the product in 95% *ee*. In 2011, this group reported an asymmetric Mannich reaction between imines formed *in situ* from aliphatic and aromatic aldehydes and amidesulphones, catalysed by prolinol 25. In these *anti*-selective reactions performed in basic solutions, β-amino aldehydes were obtained with excellent *ees* [43]. Tao's isosteviol-proline conjugates were used in water to promote three-component Mannich reactions between cyclohexanone, aromatic amines and aromatic aldehydes [44]. *Syn* Mannich products were obtained in high *drs* and *ees*. More recently, in 2013, Šebesta and co-workers developed a Mannich reaction between ethyl *N*-PMP-iminoglyoxylate and cyclohexanone which afforded β-aminoketones in a high yield of 70% and high stereoselectivity [45]. The best catalyst reported for this reaction was proline sulphonamide, which also promoted a Mannich-type cyclisation domino reaction between ethyl *N*-PMP-iminoglyoxylate and 5-hydroxypentanal. The aldehyde is available commercially as a water solution, in which it is present predominantly in the cyclic hemiacetal form, tetrahydro-2*H*-pyran-2,6-diol. The desired tetrahydropyridine was obtained in excellent *dr* (>95:5) and 96% *ee* but a low yield (20%) compared to reactions in organic solvents.

2.3. The Diels-Alder reaction

The Diels-Alder reaction was one of the first organocatalytic asymmetric reactions to be studied in pure water [46]. In 2002, Northrup and MacMillan reported that linear and cyclic enones reacted with cyclopentadiene in the presence of the chiral amine salt of oxazolidinone 26, to give bicyclic adducts in good yields and high stereoselectivities (**Scheme 2**) [47]. The reaction presumably proceeds via the formation of an intermediate iminium, a process which lowers the energy of the HOMO, facilitating the cycloaddition. When R^2 was bulky, e.g. *i*-Pr, the yield dropped considerably, and the product was racemic as a result of steric inhibition to iminium formation. In reactions of ethyl vinyl ketone and acyclic dienes, yields and *ees* were even higher and the endo selectivity was greater than 100:1 in all cases. In 2005, Lemay and

http://dx.doi.org/10.5772/64279

Ogilvie showed that aldehydes can also be used as dienophiles [48]. Protonated cyclic hydrazide 27 afforded high yields (71–96%) and *ees* (69–94%) of products in reactions of linear enals with cyclopentadiene. The diastereoselectivity was low, with either *endo* or *exo* isomers predominating according to the structure, the enal. Later, Lee described the use of camphor sulfonyl hydrazines like 28 to promote Diels-Alder reactions between aldehydes and cyclopentadiene in brine [49]. In the presence of trichloroacetic acid, high yields of predominantly endo products were obtained in high *ees*. The *dr* was once again low (1:0.9–1:2.5 *exo/endo*). In 2008, Hayashi and co-workers widened the scope of the organocatalytic aqueous Diels-Alder reaction of enals showing that a low loading of 5 mol% of TMS-protected pyrrolidine salt 29 provided high *drs* and *ees* of product with a variety of cyclic and acyclic dienes with cyclopentadiene or linear dienes [50]. Water accelerated the reaction and increased *ees* but in brine the results were worse, as a result of a detrimental salting-out effect. In this case, the reaction was *exo*-selective and pyrrolidine salt 29 afforded better *drs* than the camphor-based catalysts previously developed (62:38 to 85:15). In 2009, Xu's pyrrolidine-based catalyst 30 allowed for the first time Diels-Alder reactions of cyclohexenones with nitroolefins [51]. Water was found to provide both rate acceleration and higher *ees* with respect to organic solvents and eventually the reactions could also be performed in brine and in seawater. When the catalyst (20 mol%) was used in conjunction with 4-trifluoromethylbenzoic acid, high yields (55–99%) and *ees* (83–96%) were reported for a variety of substrates. The *exo/endo* ratio was higher than 25:1 for all the products. The dienamine obtained via reaction of the catalyst with the ketone and the Diels-Alder adduct resulting from the cycloaddition prior to enamine hydrolysis in the reaction of cyclohexenone with β-nitrostyrene could be detected by ESI-MS analysis. In 2010, Zhang and co-workers developed C_2-symmetric bipyrrolidine 31 for Diels-Alder

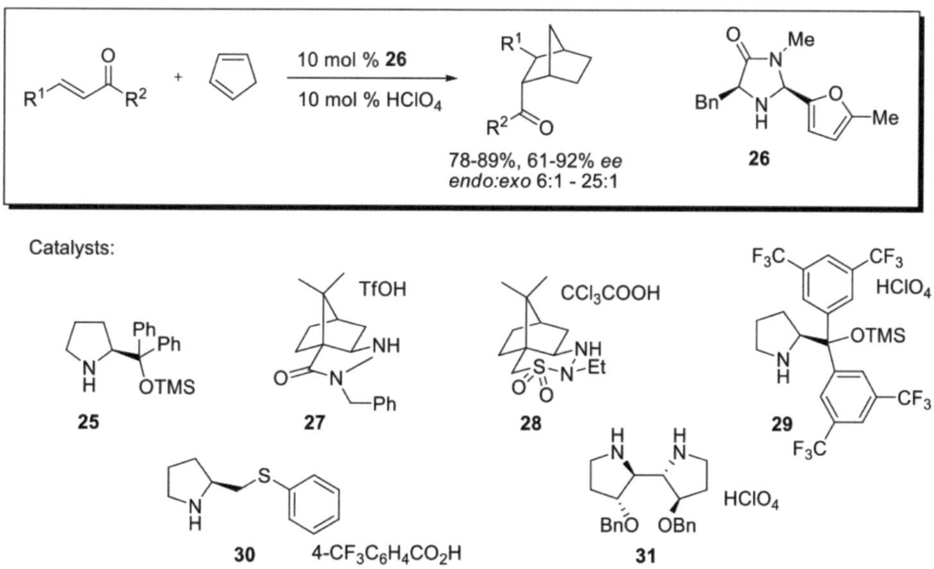

Scheme 2. The first organocatalytic asymmetric Diels-Alder reaction performed in water and other catalysts for cycloadditions in aqueous media.

reactions of enals with cyclopentadiene in water [52]. High yields of products and *ees* could be obtained but in this *exo* selective reaction *dr* was also low (0.8-1:2.2-1). The catalyst could be recycled up to five cycles without losses in stereoselectivity, but after the 4th cycle the yield dropped. In 2011, Merino showed that 1,3-dipolar cycloaddition reactions could be performed in water to afford cyclic nitrones [53].

The reaction promoted by 28 was initially developed as a one-pot procedure involving addition of aldehydes to nitroolefins, followed by *in situ* reductive cyclisation. The nitrones were obtained in good yields and more than 99% *ee*. The addition reaction was then performed on 5-hexenal. The resulting alkenyl nitrones underwent spontaneous intra-molecular 1,3-dipolar cycloaddition reactions leading to tricyclic derivatives containing four chiral in high yields and excellent *ees* simply by pH adjustment to 6 and stirring a few hours at room temperature. It is important to mention that in all cases only one enantiomer was reported in the literature.

3. Organocatalytic reaction in alternative solvents

In previous years, the search of alternative solvents as sustainable reaction media for asymmetric organocatalysis have been reported [54]. In particular, ionic liquids including chiral ones: polyethylene glycol (PEG) derivatives, organic carbonates and supercritical fluids are the major examples of alternative solvents already tested with comparable or even better performances than conventional organic solvents [55].

3.1. Ionic liquids and chiral ionic liquids

Ionic liquids as organic salts with low melting point (lower than 100°C) have emerged as environmentally benign alternative media to classic organic solvents [56]. Some peculiar properties of ILs such as their almost negligible vapour pressure, high thermal stability, high ionic conductivity, large electrochemical window, insolubility in supercritical CO_2 (*sc*CO_2) and significant dissolution performance of a large range of organic molecules and transition metal complexes are very attractive for application in organic synthesis and catalytic processes. In general, the physical and structural properties of the ILs are dependent on the suitable combination of cation/anion structures.

The possibility to use different ILs as efficient and recyclable reaction media is one important parameter for applications in organocatalysis. Additionally, the organocatalyst can be dissolved and stabilized into IL allowing to preserve its catalytic activity for several cycles [57]. In 2002, two independent reports [58, 59] showed the possibility to use ILs as alternative solvent for asymmetric aldol reaction between acetone and some aromatic aldehydes in the presence of (S)-proline (1–30 mol%) as organocatalyst. The best results (94% yield and 89% *ee*) were obtained using 1-butyl-3-methylimidazolium hexafluorophosphate, [bmim][PF_6] as IL and the organocatalyst could be reused four times without significant decrease in the yield and enantioselectivities. After these first reports, other research groups have used ILs as reaction media for different organocatalytic reactions such as Michael reaction [60]; Mannich reaction

[61]; Diels- Alder [62]; alfa-amination [63] and alfa-aminoxylation [64], among others [65]. Normally, the ILs are mainly based on alkylmethylimidazolium cation combined with BF_4, PF_6, TfO and NTf_2 anions. According some limitations related to loss of catalyst during the recovery of the product and recycle of the IL media, different approaches reported an efficient linkage of the catalyst (e.g. proline) to cationic or anionic unit from IL structure [66]. The novel class of ILs-supported catalyst have been described in the literature allowing a significant improvement of stability and recycling of the catalyst [67]. It is possible to use chiral ILs by simple incorporation of chiral organocatalyst into original cation/anion scaffold or alternatively using natural chiral cations/anions based on aminoacids (protic chiral ILs and chiral ILs based on aminoacids as anions) [68].

Chiral ionic liquids (CILs) have been recognized as having potential application for chiral discrimination, including in asymmetric synthesis and resolution of racemates [69]. A transfer of chirality in these solvents should be expected; however, only a few number of chiral ILs have been reported to date [70]. The initial report from Seddon and co-workers [71] showed the preparation of CIL [bmim][Lactate] for application in catalysis. Then, a number of new chiral ILs have been synthesized and employed as chiral additives in order to induce moderate enantioselectivity in some reactions such as in Aldol reaction, photo-isomerisation, the Baylis-Hillman reaction and Michael additions [72]. Recently, several chiral ILs have been reported based on introduction of chiral units in the organic cation or anion by efficient synthetic methods. In parallel, many examples have described the use of natural chiral sources such as aminoacids or commercially available chiral compounds such as chiral carboxylic or sulphonic acids [73]. A larger number of reported chiral ionic liquids derive their chirality from the cationic moiety. Taking advantage of the readily available chiral precursors such as amines, aminoalcohols and amino acids, it is possible to incorporate them in cationic structures.

In this context, Bica and co-workers [74] reported the synthesis and application of basic chiral ILs based on (S)-proline incorporating alkylpyrrolidinium cations and NTf_2 as anions. The authors designed these CILs in order to replace trifluoroacetic acid in enamine-based organocatalysis for asymmetric C-C bond reactions. In the case of asymmetric aldol reaction of 4-nitrobenzaldehyde and acetone, moderate to high yields and enantioselectivites (up to 80% *ee*) were obtained as a novel strategy for acid-free organocatalytic process. Then, González et al. [75] developed CILs based on chiral α-amino amides particularly derived from (S)-valine, (S)-phenylalanine and (S)-leucine for application as solvents or additives for direct enantioselective aldol reaction. Moderate to good yields and enantioselectivities for aldol reaction between p-nitrobenzaldehyde and acetone were obtained using these CILs or [bmim][NTf_2] as additives. A transfer of chirality from the chiral reaction media has been observed as well as the participation of match interactions of the chiral medium with both enantiomers of proline. The catalytic system was recovered by simple filtration, and their reuse and recycle is possible at least four times with only a slight reduction in activity. Zlotin and co-workers [76] reported a novel recyclable prolinamide-derived ionic-liquid-supported organocatalyst of asymmetric cross-aldol reactions in aqueous medium. In particular, they used aromatic aldehydes reacting with cyclic or linear ketones to give chiral aldol adducts in moderate to high yields and good *dr* (anti/syn up to 96:4) and *ee* (81–99%) values. The catalyst was recycled more than 10 times without any reduction of catalytic efficiency. Chauan et al. [77] published an efficient method for the enantioselective Diels-Alder reaction between cyclopentadiene and crotonaldehyde (94% conversion of product with exo/endo (1/1.1) and 90% *ee* of endo product) catalysed by recoverable MacMillan catalyst tailored with imidazolium ionic liquid at room temperature. In this conditions, IL-supported MacMillan catalyst is used as catalyst in the presence of trifluoroacetic acid (5 mol%) as co-catalyst, and it is reported that the catalyst can be reused up to five cycles without any significant decrease in conversions and *ee*'s values. De Nino et al. [78] described a novel chiral organocatalyst based on (5S)-2,2,3-trimethyl-5-thiobenzylmethyl-4-imidazolidinone hydrochloride for enantioselective Diels-Alder reactions in good yields with good to excellent enantioselectivities. This catalyst can be recovered and recycled for further transformations at least six times with the retention of its catalysis and enantioselectivity. In addition, only 6 mol% of catalyst and a slight excess of donor aldehyde (1.5 equiv.) are required, without additional organic solvent for final purification step. Recently, Kragl and collaborators [79] reported a novel strategy for the embedding of quinine-based organocatalysts in polymer ionic liquids-based hydrogels for application in asymmetric nitroaldol (Henry) reaction. Using this organocatalyst encapsulated into polymer IL, it is possible to recover and reuse the catalyst four times without any loss of enantioselectivity (up to 91% *ee*) and significant catalytic leaching (<0.01%).

3.2. Polyethylene glycol and deep eutectic solvents (DESs)

Polyethylene glycol as alternative media for asymmetric aldol reaction was first reported in 2004 by the Chandrasekar group [80]. The authors tested different aldehydes and ketones with comparable yields and enantioselectivities than conventional organic solvents. Also, PEG400 and Proline as catalyst were recycled at least 10 times without any decrease in the activity. In 2011, Verma et al. [81] described PEG-embedded thiourea dioxide (PEG.TUD) as an useful and recyclable host-guest complex organocatalyst for the synthesis of 3, 4-dihydropyrimidones via Biginelli condensation in order to afford the desired pure product in high yields. It is interesting to know that these results are in contrary to unreactive PEG-thiourea complexes (PEG.TU) for similar reaction condition processes. Despite the potential use of PEG derivatives as biocompatible alternative reaction media for organocatalysis, only few examples in the literature have been reported.

Deep eutectic solvents were first introduced by Abbott and co-workers [82] to describe the formation of a liquid eutectic mixture (mp 12°C) starting from two solid materials with high melting points: choline chloride (ChCl, mp 133°C) and urea (mp 302°C) in a molar ratio of 1:2. DES are generally formed by suitable combinations of two or three safe and inexpensive components which are able to engage in hydrogen-bond interactions with each other to form an eutectic mixture with a melting point lower than either of the individual components. The application of DES as alternative solvent for catalysis is very promising mainly because no purification is required; their physicochemical properties can be easily tuned according to specific reaction requirements, and they offer convenient methods of product isolation simply based on organic phase extraction or even precipitation upon addition of water, which can be subsequently removed, thereby restoring a reusable DES [83]. In recent years, DES have been applied in the fields of biotransformations, metal-catalysed reactions, organometallic chemistry and also in organocatalysis. Benaglia and co-workers [84] published three distinct stereoselective reactions (addition reactions: isobutyraldehyde to β-nitrostyrene; E-3-methyl-3-nitroethylacrylate to benzylacetone and 4-hydroxycoumarin to benzaldehyde) catalysed by a chiral primary amine through different activation methods. For these reactions, they tested three different DES (choline chloride: urea, 1:2; choline chloride: fructose; water, 1:1:1; choline chloride: glycerol, 1:2) in order to obtain the desired chiral products in high yields and enantioselectivities. Also, the use of these unconventional and biorenewable reaction media based on DES allowed the recovery and the recycling of the chiral catalyst.

3.3. Supercritical fluids and organic carbonates

The use of supercritical carbon dioxide ($scCO_2$) as an alternative medium for organocatalytic reactions is very promising from the 'green chemistry' viewpoint [85]. CO_2 in the supercritical state (critical point 31.1°C, 73.8 bar) is characterized by higher diffusion rates and a unique capability of dispersing poorly soluble reagents, thus enhancing the reaction scope, rates and selectivity. In 2014, Zlotin and co-workers [86] reported the first enantioselective organocatalytic reaction in the $scCO_2$ medium, in particular asymmetric Michael addition of diphenylphosphite to α-nitroalkenes in the presence of tertiary amine-squaramide-derived bi-functional organocatalysts. The reaction products, chiral β-nitrophosphonates, are precursors of β-amino phosphonic acid derivatives that occur in nature and possess valuable biological activities. In the reported reaction conditions (100 bar, 35°C), α-nitroalkenes enantioselectively accept diphenylphosphite in the presence of bi-functional organocatalysts bearing the tertiary amino group and the squaramide fragment to give corresponding β-nitrophosphonates in high yields and enantioselectivities (up to 94% *ee*). The use of $scCO_2$ is crucial for reaction as well as product isolation by efficient extraction and further catalyst recovery. Recently, the same group discovered the application of supercritical fluids ($scCO_2$ and $scCHF_3$) as an alternative media for asymmetric Michael domino-reactions catalysed by bi-functional tertiary chiral amines [87]. In the optimized conditions, o-N-nitrosylaminophenyl α,β-unsaturated ketones react with α-nitroalkenes in order to give functionalized chiral tetrahydroquinolenes in moderate to high yields and excellent diastereo- ($dr > 99:1$) and enantioselectivities ($ee > 98\%$).

Organic carbonates have been claimed as alternative low cost and biodegradable solvents for application in organocatalytic reactions [88]. North et al. [89] reported ethylene and propylene carbonate as an alternative solvent in asymmetric aldol reactions catalysed by (S)-proline. Using cyclic and acyclic ketones reacting with aromatic aldehydes, the desired chiral aldol products were obtained in good yields and high stereoselectivities. Additionally, an appropriate combination between propylene carbonate and the proline enantiomer was observed allowing a considerable improvement in the stereoselectivity of aldol product [90]. The same authors also reported the use of cyclic carbonates as solvents for α-hydrazination of aldehydes and ketones by diazodicarboxylates using (S)-proline as organocatalyst [91].

4. Organocatalytic reaction under solvent-free conditions

On the road to sustainability, organic chemistry has been changing and the application of catalytic processes has contributed to a more efficient use of energy, less waste and the exploration of raw materials [92]. Indeed, sustainability is a growing concern in the twenty-first century, and consequently the use of solvent-free reactions in organic chemistry is gaining importance, with a foremost impact in the environmental protection as well as on human health. Organic reactions in the absence of conventional organic solvents have become highly attractive. Consequently, over the last years, the number of reactions under solvent-free conditions has grown.

4.1. Aldol reactions

The solvent-free enantioselective organocatalysed reactions have been reviewed [93] and this chapter focuses on the most recent advances. Kumar and co-workers have recently reported two new prolinamide catalysts 32 and 33 for direct stereoselective organocatalytic and direct aldol reaction of aldehydes and ketones to produce the corresponding β-hydroxy carbonyl compounds under neat conditions [94, 95]. Catalyst based on myrtanyl-prolinamide that was synthesized in two steps from *N*-Boc-L-proline and *cis*-myrtanylamine using standard peptide coupling conditions followed by Boc group removal. The authors also explored the D-proline version of catalyst 32; however, similar results were found and only

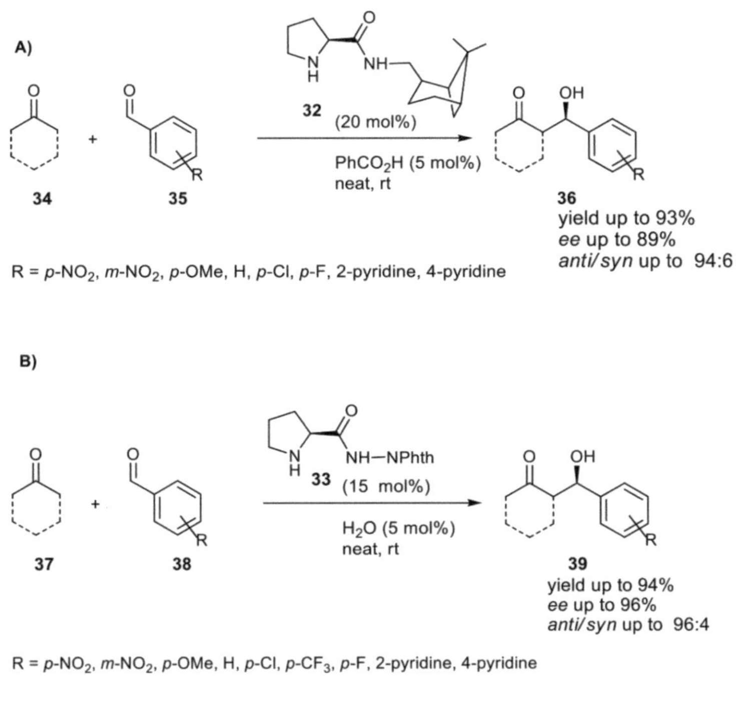

Scheme 3. Two approaches of solvent-free enantioselective aldol reactions.

this catalyst was further explored. The best catalytic conditions comprised the use of 20 mol % of catalyst and 5 mol% of benzoic acid as an additive. Several other additives were investigated such as acetic acid, formic acid, phenol, CSA, TFA and *p*-TSA. The protocol revealed to be highly effective for the preparation of several aldol adducts with high yields and high stereoselectivities under solvent-free conditions (**Scheme 3A**) [94]. Furthermore, the same authors reported the use of catalysts 33 for asymmetric direct aldol reaction, compatible with solvent-free conditions [95]. The phthalimido-prolinamide was prepared in a similar procedure, in two steps from *N*-Boc-proline and *N*-aminophthalimide. The enantioselective aldol reaction catalysed by 33 was performed using various ketones 37 and aldehydes 38 (**Scheme 3**), and proved to be effective with 15 mol% of catalyst under neat conditions. The use of 5 mol% of water accelerated the reaction and the α-hydroxy carbonyl product 39 could be attained in high yields and high stereoselectivities. Proline-containing catalysts under solvent-free conditions have also been explored by Juaristi and co-workers. They reported the use of four (*S*)-proline-containing dipeptidic organocatalysts bound to MBHA (4-methylbenzhydrylamine) resin in the asymmetric aldol reaction between cyclohexanone and several aldehydes [96]. The authors have explored different spacers (linker between the proline moiety and the resin) with different lengths in order to investigate the influence in catalytic activity in the case of spacial position of the catalyst attached to resin. Preliminary studies with all catalysts pointed 40a (10 mol%) as the best catalyst for the aldol reaction of 41 with 42 to afford 94% of isomer anti-**43** with 74:26 (*anti*/*syn*) and 77:23 (*er*). However, the conditions were optimized and the presence of benzoic acid and water as additive showed considerable improvement of the yield of 56 to 99% as well as higher stereoselectivity. Concerning the four catalysts 40a–**d**, under the optimized conditions, organocatalyst 40c gave best results. Although the results obtained were very similar for all the catalysts tested. The authors concluded that the spacer length of the bound catalyst had a slight influence on the catalytic activity, and that 40b and 40c showed an improved stereoselectivity. Furthermore, organocatalyst 40c can be reused at least for five consecutive cycles. A limitation of these catalysts is that they are not effective for less electrophilic aldehydes. Then, Jurasti and Machuca also reported the use of a dipeptidic organocatalyst in the asymmetric aldol reaction between 41 and several aldehydes under solvent-free conditions and mechanochemical activation in a ball mill [97]. Previous studies suggested that the catalysis by dipeptides and prolinamides is operative [98]. The authors proposed that the pyrrolidine moiety in catalyst activates the ketone through formation of a chiral enamine intermediate, and the aldehyde is activated by the formation of a strong hydrogen bond, via amide NH. Furthermore, they proposed that a non-covalent π-π interaction between aromatic rings of the catalyst and the aldehydes leading to a rigid transition state and inducing a higher stereoselectivity in the condensation reaction. Thus, the authors decided to explore the catalyst containing a naphthyl substituent in order to obtain further data to support the proposed non-covalent π-π interaction between the aromatic ring of catalyst and aldehyde units. The fact that the reaction is carried under solvent-free conditions and with the use of HSBM (high speed ball milling) allowed reduction of the molecular motion. Thus, this dipeptide catalyst was prepared via condensation of *N*-Cbz-(*S*)-proline and the hydrochloric salt of the methyl ester of (*S*)-naphthylalanine and subsequent *N*-deprotection via hydrogenation under Pd/C. Several aromatic aldehydes were tested and the reaction was carried in a ball mill. Higher stereoselectivities and higher yields (except for aldehyde possessing *m*-Cl group) were obtained with aldehydes possessing electron-withdrawing groups on the aromatic ring. Best stereoselectivity was reported for aldehydes possessing a *p*-metoxy and *per*-fluorinated aromatic ring (99:1, *anti/syn*) and *er* (99:1). The results obtained supported the proposed transition state in which a π-stacking between naphthyl ring of the catalyst and electron poor aromatic aldehyde furnishes a more rigid transition state and therefore a higher stereoselective aldol reaction.

Additionally, the same group has further used solvent-free conditions and evaluated the use of three (*S*)-proline containing dipeptides as organocatalysts in asymmetric aldol reactions of cyclohexanone (41) p-nitrobenzaldehyde (42) and under ball-milling conditions [99]. The authors have used three different protocols: solvent-free under HSBM activation; neat conditions with conventional stirring; and in solution phase. The HSBM was the best strategy since shorter reaction times were needed and higher stereoselectivities were observed. Some catalysts led to comparable yields and stereoselectivity [98% yield, 90:10 (*anti/syn*) and 90:10 *er*]. Thus, the authors concluded that the presence of a second stereogenic centre in the α,β-dipeptide organocatalysts did not result in a higher stereoselectivity comparing with examples with one stereogenic centre. Asymmetric aldol reactions under solvent-free conditions have been further explored by Nájera and Gómez-Bengoa [100] comparing with examples with one stereogenic centre (**Scheme 4**).

4.2. Michael addition reactions

Asymmetric organocatalysed reactions under solvent-free conditions have been extended to Michael addition reactions. Recently, Bolm and co-workers developed an efficient, solvent-free protocol for Michael addition reactions of α-nitrocyclohexanone to nitroalkenes using thiourea derivatives as catalysts [101]. These reactions have been carried out in sustainable conditions, using planetary ball mill, with low catalyst loading and short reaction times, leading to high yields (up to 97%) and high enantioselectivities (*er* up to 98:2). In a previous study, the authors have developed the use of natural amino acid-derived bi-functional thioureas as bi-functional organocatalysts in the asymmetric Michael addition reaction [102]. On the follow-up of this work, the authors investigated mechanochemical effects induced by ball milling on the aforementioned thiourea-catalysed asymmetric Michael addition reactions of α-nitrocyclohexanone to nitroalkenes to provide desired products. The reaction conditions were optimized such as the catalyst loading. The reaction proceeded with high stereoselectivity even with 1 mol% of catalyst; however, the yield was lower (57%) and thus 2.5 mol% of catalyst was used. Thiourea-organocatalysts have also been explored by Hestericová and Šebesta as hydrogen-bonding organocatalysts for the Michael addition reaction to nitrostyrene under solvent-free conditions [103]. Thus, different thioureas have been used involving aminoalcohols, Cinchona alkaloid, binaphthyl diamine, bis(thiourea) and also squaramides. The authors have performed comparative reactions in a ball mill and in solution. In terms of yields, both reactions proceeded well; however, a large variation of yields was obtained (19–95%) under solvent-free conditions. The use of an additive such as benzoic acid or mandelic acid (20 mol%) improved the yield from 50 to 80% and 95%, respectively, in the case of the reaction carried in the ball mill. Concerning enantioselectivities, the best results were obtained with thiourea catalysts as well as with Cinchona-containing catalyst, which increased the enantiomer ratio up to 10:90. A pyrrolidine-diaminomethylenelononitrile organocatalyst was developed by Miura and co-workers for the Michael addition of carbonyl compounds to nitroalkenes under solvent-free conditions [104]. The novel organocatalyst promotes the asymmetric conjugate addition of cyclohexanone to nitroalkene to afford the corresponding adduct in high yield with up to 99% *ee*, under solvent-free conditions. The novel organocatalyst based on the skeleton of DMM (diaminomethylenemalononitrile) efficiently catalysed the reaction employing low amounts of catalyst and in a short reaction time under mild conditions, affording high enantioselectivity. The authors proposed that the DMM skeleton can act as an efficient double hydrogen bond donor for Michael additions to nitroalkenes under solvent-free conditions. A highly efficient and simple asymmetric organocatalytic Michael addition of α,β-disubstituted aldehydes 44 to nitroolefins 45 under solvent-free conditions was developed by Ni and co-workers [105]. The authors have developed a chiral pyrrolide-based diamine 46, that in combination with benzoic acid, proved to be very effective in order to prepare the adduct 47 in high yields and high enentioselectivities (up to 96% *ee*) with a wide range of Michael acceptors (**Scheme 5**).

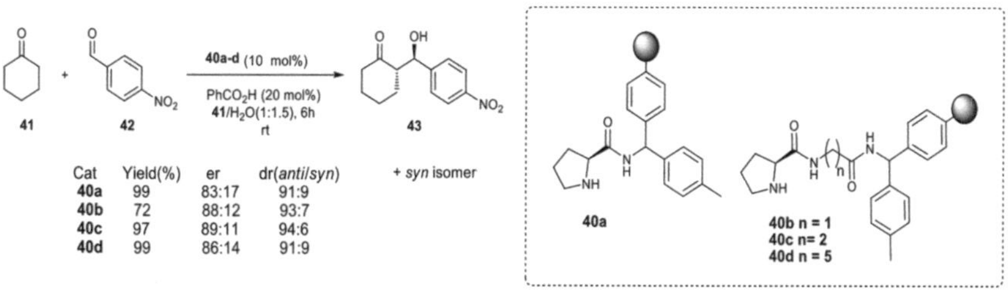

Scheme 4. Solvent-free approach using (S)-proline containing dipeptides as organocatalysts in asymmetric aldol reactions.

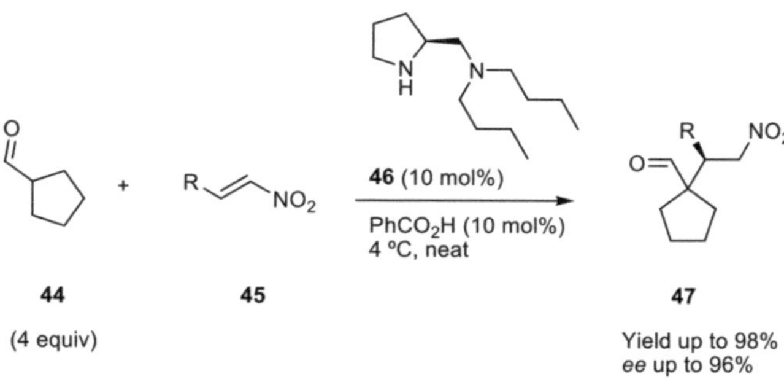

Scheme 5. Highly efficient and simple asymmetric organocatalytic Michael addition of α, β-disubstituted aldehydes to nitroolefins under solvent-free conditions.

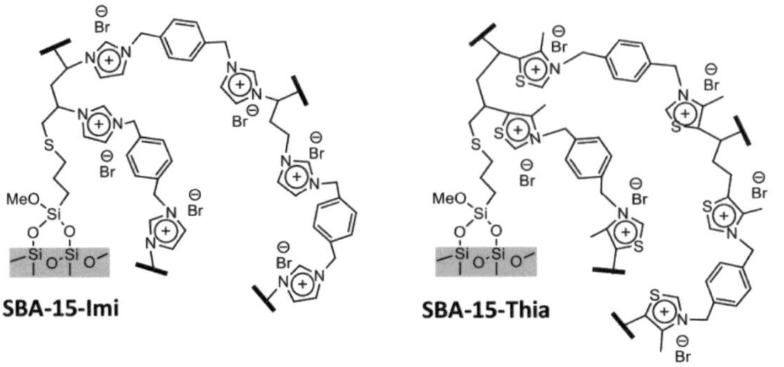

Scheme 6. SBA-15 mesoporous silica functionalized with imidazolium (SBA-15-Imi) and thiazolium groups (SBA-15-Thia).

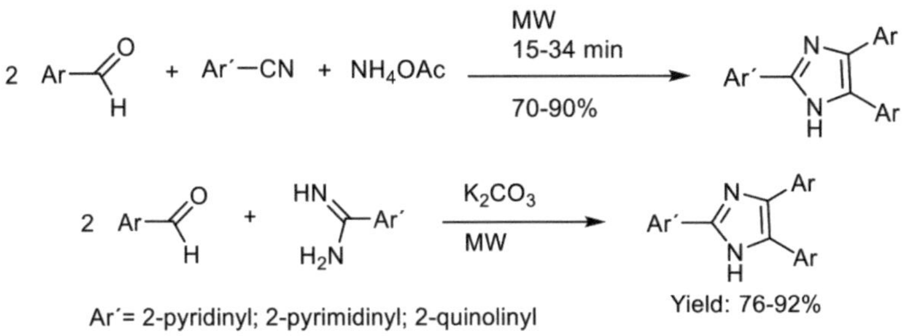

Scheme 7. Multi-component synthesis of 2-(2′-azaaryl)imidazoles under microwave irradiation.

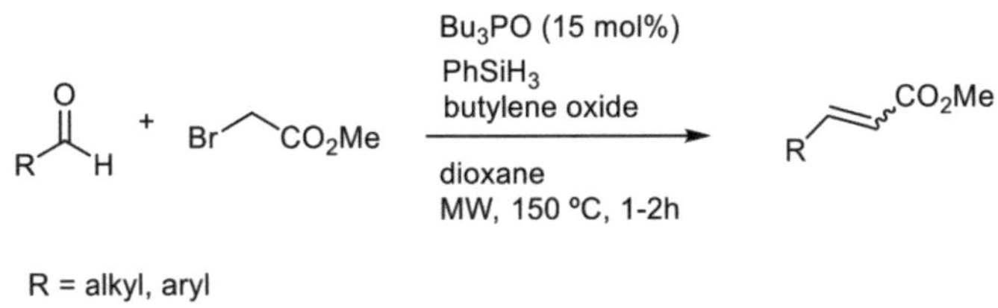

Scheme 8. Microwave-assisted catalytic Wittig reaction.

The organocatalytic Michael reaction of ketones with γ-monohalonitrodienes was reported by the Xu group using chiral prolinethiol ether as organocatalyst for the synthesis of functionalized monohaloalkenes, under solvent-free conditions [106]. After optimisation of the reaction conditions, the reaction scope was examined and several substitutions on the aromatic ring were investigated (for X = Br), and several groups were well tolerated (e.g. *p*-Cl and *p*-CF_3), as well as several ketones. Furthermore, the enantioselectivity of the reaction was not much affected by the substituents on the cyclic ketone. The authors highlighted that the reactions proceeded on the sterically less hindered carbon, which is opposite to the other acyclic ketones. They propose that the balance between steric effects and stability of the enamine derived from the catalyst and ketones might favour the formation of the terminal Michael adduct. Different substituents on the γ-position of the nitryl scaffold demonstrated that nucleophiles possessing substituents can be used to form the desired adducts in high diastereoselectivities and high to excellent enantioselectivities. The method represents a novel approach for accessing highly functionalized monohaloalkenes with α,β stereocentres of up to >99% *ee*. In 2015, Jurasti and co-workers reported the organocatalysed Michael addition reaction under solvent-free conditions of ketones to nitro-olefins [107]. The authors synthesized several organocatalysts derived from chiral (*S*)-proline containing a thiohydantoin moiety. The prepared thiohydantoins were evaluated as organocatalysts in asymmetric Michael addition under solvent-free conditions, using cyclohexanone and β-nitrostyrene as models.

4.3. Mannich reaction

Organocatalysed Mannich reactions have also been recently explored under solvent-free conditions. Fioravanti and co-workers have reported the synthesis of trifluoromethyl *syn-* or *anti*-amino alcohols by one-pot solvent-free Mannich-type reactions under temperature control [108]. On the basis of their previous results [109], these authors explored a one-pot solvent-free approach [110]. Thus, a L-proline catalysed Mannich-type reaction was investigated to prepare nitrogen-containing organofluorine compounds. On the expectation that the presence of a stereocentre in the α-position to the electrophilic site of not isolated (R,E)-aldimines could influence the diastereoselective reaction outcome, the authors started from the chiral amines(*R*)-1-phenylethylamine and (*R*)-1-(*p*-methoxyphenyl) ethylamine. The results reported demonstrated that at room temperature a mixture of *syn*/*anti* was always formed, while changing the temperature only *syn* or *anti* isomers were formed at 40 and 0°C, respectively. The products were detected by ^{19}F NMR experiments of the crude mixtures. Moreover, the chiral stereocentre on the benzyl moiety strongly affected the stereoselectivity of the reaction. The absolute configurations of the created chiral centres were determined. The protocol developed by Fioravanti and co-workers consists on a highly diastereoselective one-pot solvent-free synthesis of fluorinated *syn-* or *anti*-amino alcohols by an environmental friendly approach. Under solvent-free conditions, the authors highlight the presence of a resident stereocentre in the α-position to the imine carbon that leads to the facial stereoselective control of nucleophilic attack, suggesting that the proline is able to control only the *syn* or *anti* diastereoselectivity. A theoretical study was developed by Parasuk and co-workers about the factors that influence stereoselectivity in proline-catalysed Mannich reactions [111].

5. Organocatalytic reaction using heterogeneous systems

The immobilisation of homogeneous organocatalysts using several supports has been quite explored in previous years since in general their heterogenisation allows more stable and efficient catalyst. These parameters are aligned to the demands of sustainability and economical scalability issues. Several supports have been used: mesoporous silica [112, 113], biopolymers as chitosan [114, 115], synthetic polymers as polystyrene and polyacrylamide [116], carbon nitrides [117], metal organic frameworks (MOFs) [118], dendrimers [119], graphene [120] and magnetic nanoparticles [121–123].

5.1. Recent approaches in heterogeneous organocatalysts

Corma and Garcia have reported silica-bound organocatalysts as heterogeneous, recoverable and recyclable catalysts in several organic transformations [112]. Heterogeneous organocatalysts based on organically modified hybrid mesoporous silica (mainly MCM-41 and SBA-15) and their efficiency in several organic transformations have been also reviewed by Rostamnia [113]. These types of supports are very stable, biocompatible and can be functionalized with a wide range of functional groups. In general, their resultant-supported organocatalyst is more stereoselective, chemoselective and efficient than the homogeneous analogous. The organic moieties supported include: amines (primary, secondary and tertiary), sulphonic acids, acid-based bi-functionalized systems, ephedrine, proline, urea, thiourea and guanidine and fluorinated alcohol [114]. Kadib [114] and Mahé et al. [115] summarized the field of organocatalytic reactions promoted by chitosan used as an insoluble organocatalyst or as a support for organocatalysts. Chitosan is ranked as the second most abundant polysaccharide after cellulose and it is obtained from deacetylation of chitin, which is exclusively extracted from industrial marine discharge. Chitin is constituted by *N*-acetyl-d-glucosamine monomers connected through β (1–>4) linkages. The presence of amino groups on the polymer backbone enables the covalent linkage of different functional groups to its skeleton. Several organocatalysts based on pyridine, proline, ionic liquids and quaternary ammonium have been supported. These chitosan derivatives showed reactivity similar to their homogeneous analogous, nevertheless the introduction of cooperative acid-based interactions enhances significantly their reactivity. Carbon nitride, for simplicity C_3N_4, mainly composed of C and N, is one of the oldest reported polymers and has many advantages when compared with traditional heterogeneous catalysts, it is abundant and cheap, metal free, tunable electronic structure and has good thermal and chemical stabilities. The group of Antonietti [117] has emphasized the recent breakthroughs in their modification and their applications as sustainable catalysts in several reactions, such as photochemical splitting of water, mild and selective oxidation and hydrogenation reactions, and in photodegradation of pollutants.

In the previous years, dendrimers have also attracted the attention of the scientific community as they combine the advantages of homogeneous catalysts, showing fast kinetic behaviour, and heterogeneous catalaysts, since they can be easily separated from the reaction mixture by precipitation, membrane or nanofiltration methods. Wang et al. described the recent advances for metallodendritic catalysts and dendritic organocatalysts [119]. Magnetic nanoparticles are another interesting supports for heterogenisation of organocatalysts since it allows their recovery with sustainable techniques of magnetic separation. Magnetite, also known as ferrite (Fe_3O_4), has a very active surface suitable for functionalisation or adsorption of several metal- and organic-based catalysts. In general, these heterogeneous catalysts are highly stable and can operate under mild conditions, using environmentally benign solvents or even water, with good performances and recyclabilities [121–123]. They have been applied in a wide range of reactions, such as Mannich-type reactions, C-C, C-S and C-O coupling reactions, alkylation, oxidation, reductions and asymmetric synthesis. Mrówczyński et al. summarized their use as supports for organocatalysts [122], their use in asymmetric catalysis has been reviewed by Dalpozzo [123] and their application in catalysis, green chemistry and pharmaceuticals reactions are described by Gawande et al. [121]. Bartók reports the advancements of heterogeneous asymmetric direct aldol reactions using organocatalysts based on hydroxyproline, prolinamide and peptides immobilized by covalent or ionic bonding and by adsorption on different supports [124]. In order to allow the application of organocatalysts in industry, their scale-up using continuous flow technology has attracted much attention in recent years. Three very interested reviews in this field were published recently [125–127]. Puglisi et al. [126] and Atodiresei et al. [127] reported several types of asymmetric organocatalysed reactions in continuous flow and highlighted their advantages over batch reactors. Heterogeneous-supported organocatalysts are focused on both reviews. This field is still in its infancy since the examples known are applicable mostly for particular substrates and with some problems of catalyst deactivation. However, in general these processes improve the efficiency of the organic transformations by reducing the amount of catalyst loading and reactions times. Very recently, Munirathinam et al. [125] reviewed the main achievements that has been in this area but focusing on a broader range of supported catalysts including acid, base, organomettalic, peptidic, enzymatic, ionic liquids and metal nanoparticles by using the three main approaches to incorporate them into the catalytic micro-reactors: (i) packed-bed, (ii) monolithic and (iii) inner wall-functionalized. The application of these catalytic micro-reactors on several reactions and their advantages over classical batch reactors were also presented.

5.2. Some examples of heterogeneous organocatalysts in solvent-free conditions

SBA-15 mesoporous silica functionalized with mercaptopropyl groups were used for the covalent immobilisation of multi-layered ionic-liquid-like phases containing imidazolium or thiazolium active sites [128]. These new hybrid materials were used for the etherification of 1-phenylethanol under solvent-free conditions at 160°C under different gas phase (oxygen, air, nitrogen and argon). The best catalytic performances were obtained for the material bearing thiazolium groups under oxygen, and this hybrid material also showed higher catalytic activity (92% of conversion and 75% of selectivity, under O_2, 160°C, 7 h) when compared with its homogeneous analogous catalyst (92% of conversion and 72% of selectivity in the same reaction conditions). For example, the heterogeneous catalyst was recycled seven times, without loss of activity, for the etherification of 1-phenylethanol. Other two alcohols were also tested: benzyl alcohol and diphenylmethanol. García-Suárez et al. [129] have tested for the first time the catalytic activity of a Bio-IL [Chol][Pro] (choline-proline) in the Michael addition reaction and supported this catalyst on different heat-treated mesoporous carbon materials by simple physical adsorption in organic media, reporting also the catalytic activity of the heterogeneous systems. The coupling of cyclohexanone and β-nitrostyrene to produce 2-(2-nitro-1-phenylethyl)cyclohexanone was selected to evaluate the catalytic activity of the catalysts, under solvent-free and at room temperature conditions. Excellent conversions and high diasteroselectivities were obtained for the heterogeneous catalysts based on commercially available mesoporous carbon beads heated at 1500°C and 2000°C. These results are similar to those obtained for the homogeneous catalyst. The stability of the supported Bio-IL is strongly influenced by the textural and surface chemical properties of the supports tested.

Recently, the group of Wang [130] described a new method for the hollow-structured phenylene-bridged periodic mesoporous organosilica (PMO) spheres using hematite (α-Fe_2O_3) nanoparticles as a hard template. These materials were functionalized with MacMillan catalyst (H-PhPMO-Mac) by a co-condensation process and a 'click chemistry' post-modification and by grafting. For comparison, analogous materials were prepared in the absence of hematite. Their catalytic activity was tested in asymmetric Diels-Alder reaction using water as solvent. The model reaction tested was the Diels-Alder cycloaddition of 1,3-cyclopentadiene with trans-cinnamaldehyde. The catalyst H-PhPMO-Mac has shown higher catalytic activity (98% yield, 81% enantiomeric excess (*ee*) for endo and 81% *ee* for exo) in water than its non-hollow analogous and it can be reused for seven runs without a significant loss of activity. For the homogeneous catalyst in the same conditions, lower yield of product (80%) and higher *ee* for endo (93%) and exo (91%) were obtained. The materials functionalized with the catalyst by grafting exhibit lower catalytic efficiency. Colloidal graphene oxide was synthesized through a modified Hummer's method, avoiding additional hazardous treatments, and its activity as base catalyst for the condensation of several substituted benzaldehydes with acetophenone (aldol reaction) and with active methylene compound malononitrile (Knoevenagel reaction) at room temperature and under solvent-free conditions was tested [131]. The heterogeneous organocatalyst showed high reactivity in 8 h toward Knoevenagel condensation with 97% of conversion and 99% of product selectivity. It can also be reused for five cycles without any loss of activity (**Scheme 6**).

The same group reported for the first time the incorporation of 4-(*N,N*-dimethylamino)pyridine (DMAP) into the network of a nanoporous-conjugated polymer (NCP) prepared through the Sonogashira-Hagihara coupling reaction of rigid building blocks of DMAP monomer and a structural linker [132]. A nanoporous structure, mainly with mesoporous, with highly concentrated and homogeneously distributed DMAP catalytic sites was obtained and its activity in the acylation of alcohols studied. This catalyst has shown excellent catalytic performance for the conversion of several aliphatic alcohols and phenols into the corresponding ester products, at room temperature and with dry dichloromethane as solvent (if necessary), with yields higher than 90%. This heterogeneous catalyst can be reused for 14 cycles without significant loss of activity and run for 536 h under continuous-flow conditions, which shows its potential as robust heterogeneous catalyst for industrial use [132]. Guan et al. prepared bi-functionalized SBA-15 and Al-SBA-15 mesoporous materials with different matches of acid and base by immobilisation of different organic amines, 3-aminopropyltriethoxysilane and 3-(triethoxysilyl)propylcarbamoyl)pyrrolidine, through a post-synthesis method. The materials were tested in a Knoevenagel reaction, tandem deacetalisation-Knoevenagel reaction, one-pot deacetalisation-Henry reaction, aldol reaction and nitroaldol reaction. In these systems, there is a synergic effect between the acid and base which favoured different reactions. For example, weak acid matching weak base favoured the Knoevenagel reaction and nitroaldol reaction, while moderately strong acid matching weak base showed good results for one-pot deacetalisation-Knoevenagel reaction, one-pot deacetalisation-Henry reaction and aldol reaction [133]. A bi-functional catalyst was prepared by covalent immobilisation of trans-4-hydroxy-L-proline on graphene oxide via a succinate spacer and its catalytic performance tested for the solvent-free ketene forming reaction of benzaldehyde and acetone [134]. The results demonstrated that it behaves as an efficient, recoverable and recyclable catalyst (for five cycles). The effect of different solvents on its performance was also studied.

A new approach to obtain chiral metal organic frameworks as heterogeneous asymmetric photocatalysts through the cooperative combination of stereoselective organocatalyst L-or D-pyrrolidin-2-ylimidazole (PYI) and a triphenylamine photoredox group into a single framework was developed by Duan, He and co-authors [135]. Two enantiomeric MOFs of Zn were prepared and applied to prompt the light-driven α-alkylation of aliphatic aldehydes with high catalytic efficiency and enantioselectivity. For comparison, lanthanide-based MOFsHo-TCA (H3TCA = 4, 4′,4′′- tricarboxyltrihexylamine and MOF-150, assembled from 4,4′,4″-nitrilotribenzoic acid, were studied and the results suggested that both photosensitizer triphenylamine and the chiral organocatalyst were necessary for the light-driven reaction. However, the corresponding MOF obtained by mixing the chiral moiety has shown lower enantioselectivity. Two chiral porous MOFs functionalized with carboxylic acid groups were reported for the first time by Liu et al. [136]. One of them was able to encapsulate S)-2-(dimethylaminomethyl) pyrrolidine by combining the carboxylic acids and chiral amines *in situ* through acid-base interactions. This organocatalyst revealed to be an efficient and recyclable heterogeneous catalyst for the asymmetric direct aldol reactions of both acetone and cyclohexanone with nitro-substituted aromatic aldehydes, in a ketone/water mixture at room temperature. The results indicated a significant enhancement of stereoselectivity by comparison with its homogeneous organocatalyst. The yield/*ee* values for the three consecutive runs were 73/74, 72/73 and 70/73%, respectively. Krishnan et al. [137] described for the first time the use of polystyrene-supported poly(amidoamine) (PAMAM) dendrimers of first, second and third generations as highly efficient and heterogeneous basic organocatalyst in Knoevenagel condensations of carbonyl compounds with active methylene compounds. The third generation catalyst gave the better results. The reactions of several carbonyl compounds proceeds very well in ethanol for short periods of times, at 30°C or 50°C, with product yield in the range of 95–100%. The catalyst can be recycled up to 10 times with only a decrease in the product yield from 100 to 97% and an increase of time reaction from 15 to 20 min for the reaction of benzaldehyde with malonitrile at 30°C. This catalyst is also environmentally friendly, once the use of aromatic and halogenated solvents and complicated purification processes is avoided.

6. Organocatalytic reactions using sustainable synthetic protocols

Over the past two decades, many efforts have been made both in industry and academia to develop synthetic organic protocols using more efficient methodologies with the aim to protect the environment and prevent waste. With that goal in mind, sustainable mechanochemical processes such as high-speed ball [138–140] microwave (MW) [138, 141, 142] or ultrasound

[143] are being increasingly used in the synthetic organic chemistry [144]. Employing this unconventional energy inputs, it is conceivable to offer innovative and highly appropriate alternatives to traditional synthetic processes [145, 146]. The use of microwave or ultrasound in organocatalytic processes have been previously reviewed by others [147, 148].

MW irradiation offers several advantages over conventional heating, such as instantaneous and rapid heating (deep-inside heating), high temperature homogeneity and selective heating [138, 142, 148, 149]. The observed enhancement of the reaction rate is in part associated with the rapid heating caused by MW irradiation relative to the same reaction using conventional heating. Beyond the controversial debate around the existence or not of non-thermal microwave effects, we must have to accept that MW chemistry is an effective, safe, rapid and highly reproducible way to perform chemical reactions that recently were translated to continuous flow processes [149, 150]. Also, the possibility to performing the MW and ultrasound reactions in the absence of a solvent is a major advantage [151]. An example is the Michael addiction of diethyl malonate to several enones catalysed by (*S*)-proline (15 mol%) under solvent-free conditions. The Michael adducts were obtained in short reaction times with good yields and moderate to excellent enantioselectivities (40–99% *ee*) [152]. In another procedure, pyranones were prepared under microwave irradiation and solvent-free conditions in ball-milling through a oxa-Diels-Alder reaction between α,β-unsaturated ketones and aldehydes. Enantioselectivities up to *e.r.* 63:37 were observed with chiral pyrrolidine-based organocatalysts [153]. Also under solvent-free conditions, Qaroush and collaborators reported a microwave-assisted preparation of [6]-oligourea based on an isocyanate-free method utilising propylene carbonate as a green carbonylating agent and 1,6-hexamethylene diamine under catalysis of 1,5,7-triazabicyclo[4,4,0]dec-5-ene (TBD). Whereas dynamic mode microwave-assisted synthesis gave 79% yields of [6]-oligourea, almost quantitative yields were obtained using the fixed mode, within 20 min, at 10 W and with the same catalyst loading [154]. The imidazole ring is a very important motif since this ring is the key constituent of a range of bioactive compounds [155]. Multicomponent approaches between aromatic aldehydes and heteroaryl nitriles under solvent-free and microwave-irradiation conditions led to highly substituted 2-(2'-aza-aryl)imidazoles. *Anti*-1,2-diarylethylbenzamides could also be obtained in an efficient six-component approach between aldehydes and ammonium acetate [156]. The syntheses were finished within short periods (15–34 min) with good to excellent chemical yields and stereoselectivity. An interesting mechanistic approach was proposed by the authors for these two reaction processes [156] (**Scheme 7**).

Recently, the same authors describe an efficient three-component domino [3+1+1] heterocyclisation to 2-(2′-aza-aryl)imidazoles promoted by K_2CO_3 under microwave irradiation conditions [157]. This one-pot operation makes use of mild conditions and short reaction times of 20–32 min and excellent atom economy. An interesting mechanism involving a umpolung process has been proposed for the formation of the 2-(2′aza-aryl)imidazoles [157]. Catalytic amounts (10 mol%) of bis-arylureas and bis-thioureas promote the Friedel-Crafts alkylation between nitroolefins and aromatic and heteroaromatic N-containing derivatives [158]. Best results are noticed on running the reactions in the absence of solvent. When applied to indoles, this protocol provides the corresponding Michael adducts in good to excellent yields and with high selectivity. *L*-Proline is actually one of the most studied catalysts in organic catalytic transformations [159], in particular involving Michael additions [160], and as expected quite applied also in transformations using unconventional energy inputs. The *L*-proline catalysed Michael addition of aldehydes and ketones to *trans*-nitrostyrene using microwave irradiation was investigated by Russo et al. [161]. High yields, short reaction times and comparable diastereo- and enantioselectivity were obtained under simple and more environmentally benign conditions such as the use of ethanol as solvent and only a slight excess of the carbonyl compound. Omara et al. [162] using microwave-assisted technique and *L*-proline catalysed the Michael addition reaction of aldehyde to α-styrene using the ionic liquid 1-butyl-3-methylimidazolium bis(trifluoromethylsulfonyl)inide ([bmim]NTf$_2$) as reaction medium. The Michael product was obtained in excellent yield (96.5%) and 36.9% *ee* with 10 mol% catalyst loading, 5.0 min reaction time and 2.0 substrate equivalent ratio. Other amines where some natural alkaloids are included are also used as catalyst in organic Michael additions. A quadruple cascade Michael/Henry condensation/Michael/aldol condensation reaction allows an efficient asymmetric synthesis of tri-substituted cyclohexene carbaldehydes employing acetaldehyde and nitroalkenes as substrates. Moderate to good yields (25–45%) and high enantioselectivities (*ee* = 89–99%) were obtained [163]. Enders et al. performed the reaction between acetaldehyde and nitrostyrene in dioxane at room temperature using (*S*)-diphenylprolinol TMS-ether (20 mol%) as a catalyst. This procedure has the disadvantage of needing 14 days to complete affording the aldehyde with a moderate yield, high enantioselectivity and a good diastereomeric ratio. In an attempt to try to improve the results, the authors observed that the presence of water accelerates the reaction substantially and also tested the reaction under microwave irradiation [164]. *L*-Proline and proline analogues were used as catalyst for the α-amination of di-substituted aldehydes with azodicarboxylates under microwave conditions. It could be observed that enantioselectivity and yield could be significantly increased by the use of microwave irradiation. Although the catalyst loading is slightly high, this is an interesting protocol that allows the α-amination of branched aldehydes. An extensive study was performed by the authors changing several parameters as the temperature, solvent and reaction time [165]. Chiral squaramides have proven to be very effective hydrogen-bonding organocatalysts with application in several asymmetric transformations [166]. Sánchez-Roselló and co-workers applied chiral squaramides in intra-molecular aza-Michael reactions under microwave irradiation with conjugated *N*-acyl pyrazoles as ester-type Michael acceptors. This protocol was especially efficient in the formation of piperidine derivatives and the authors evaluate the synthesis of different 6-memberering heterocycles [167].

Chiral tetrahydroisoquinoline-based guanidines were prepared by Naicker and collaborators [168] using a microwave-assisted synthesis. The prepared catalyst was applied as chiral catalysts to promote the asymmetric 1,4-addition of β-keto esters or malonates to nitroolefins in up to 97% *ee*. α-Spirolactones and α-spirolactams can be obtained in an overall transformation involving an olefin cross-metathesis followed by an intra-molecular organocatalytic Michael-induced spirocyclisation under microwave irradiation. The Hoveyda-Grubbs catalyst can be used on the metathesis reaction and also as a spirocyclisation *N*-heterocyclic carbene catalyst [169]. Thiazolium derive *N*-heterocyclic carbene (NHC)-catalysed

Stetter reaction of acetyl anions to various α, β-unsaturated acceptors under microwave irradiation. This procedure involving microwave heating significantly increased the chemical efficiency by the substantial reduction of reaction time [169].

The aldol reaction is one of the most important carbon-carbon bond formation reactions widely employed in synthetic organic chemistry. Proline has also an important role as organocatalyst on this reaction. Liao and co-workers [170] developed a microwave-based procedure to promoted direct aldol condensation using polystyrene-supported amine catalyst. Microwave greatly shorten the reaction times to only 20 min and improved the yield significantly. This procedure has the advantage of recovering the catalyst by simple filtration and can be reused for at least four times without significant loss of reactivity. The synthesis of heterocycles can be achieved under mild conditions using microwave irradiation. Gangwar et al. [171] using oxalic acid as catalyst reported the preparation of 3,4-dihydropyrimidin-2(1H)-one derivatives by Biginelli reaction between aromatic aldehydes, ethylacetoacetate or methylacetoacetate and urea under microwave irradiation for 2–5 min. The oxalic acid was used in very low quantity (2 mol%). The antioxidant properties were evaluated and the compounds having –OH group on benzene ring were found to have higher activity. A new, efficient and convenient approach to the synthesis of new extended angular fused aza-heterocycles including dibenzacridine and naphth[2,3-a:2',3'-j] acridine units with good luminescent properties is described [172]. The multicomponent reactions (MCRs) were conducted by reacting readily available and inexpensive starting materials using thiosalicylic acid as a catalyst under microwave irradiation. A total of 14 examples were examined, and a broad substrate scope and high overall yields (72–89%) were revealed. 1,2,3-Triazoles are an interesting class of heterocyclic unit widely used in the discovery and modulation of drug candidates. The copper-free cycloaddition reaction of azidophenyl arylselenides with β-keto-esters under catalysis of diethylamine and microwave irradiation allowed the synthesis of high-functionalized-1,2,3-triazole in good to excellent yield. With microwave irradiation it was possible to reduce the reaction time from hours to few minutes [173]. The proposed mechanism involves [3 + 2] cycloaddition reaction between the azide group and the enamine followed by elimination of the diethylamine catalyst. A very interesting MW-assisted formation of polysubstituted salicylaldehydes from propargyl vinyl ethers using imidazole as catalyst was developed by Tejedor et al. [174]. A diverse array of salicylaldehydes from simple aromatic monocyclic to complex fused polycyclic systems was obtained in moderate to high yields (38–72%). Using this procedure, it was possible to achieve the benzophenone-derived natural product morintrifolin B in a five-step synthesis. The authors proved that the reaction is scalable and instrumentally simple to perform, highly regioselective and takes place under symmetry-breaking conditions. Symmetrically substituted propargyl vinyl ethers afforded asymmetrically substituted salicylaldehydes. Some protocols are being developed for the microwave-assisted catalytic Wittig reaction. Recently, Hoffmann, Werner and Deshmukh [175–177] address this subject and a very extensive study was carried out to find the scope and limitations of this reaction. Among the several catalysts tested, epoxides proved to be suitable masked bases for this reaction. Phosphine oxides $Bu_3P = O$ proved to be the most promising catalyst that can be reduced *in situ* with silanes to generate Bu_3P as the actual catalyst. Good isolated yields and excellent *E*/*Z* selectivities were achieved. In respect to the halide component, 2-bromoacetonitrile proved to be particularly suitable, giving the desired product in yields of up to 88%. Using chiral bis-phosphines enantiomerically enriched alkene with a *er* of 81:19 were obtained (**Scheme 8**).

In another approach, McNulty and collaborators [178] have shown that it is possible to achieve high (*E*)-olefin selectivity employing phosphonium salts in water as the solvent and with weak bases, including secondary amine catalysis under MW irradiation. Extension of this procedure for the preparation of stilbenes under physiological conditions is also described. A very simple work up of the reaction involves filtration or aqueous organic partition which allows easy separation of water-soluble phosphine oxides. Taking advantage of the higher temperature achieved via either microwave (vessels heated to 190°C over 5 min) or conventional heating (5–40 min at 190°C), the decarboxylation of L-histidine and other L-amino acids were achieved with R-carvone as catalyst [179]. The use of carvone can be of advantage as any unrecovered R-carvone catalysts can hydrolyse at high-temperature to carvacrol. Ultrasound is considered an environmentally clean technology, which is based on the application of sound energy. With this technology it is possible to speed dissolution by breaking intermolecular interaction and provide the energy for certain chemical reactions to proceed. This can lead to an increased yield and selectivity of the products avoiding adverse reactions conditions that use long reaction times and high temperatures [180]. Mangilal et al. reported [169, 181] the regioselective ring opening of *trans* spiro-epoxyoxindoles with aniline derivatives, from the less hindered end, to obtain β-hydroxy-β-amino esters, in water under sonication. From the several organocatalysts used, quinine combined with urea-hydrogen peroxide (UHP) was proved to be the best for the diastereoselective epoxidation of (*E*)-3-ylidene-indolin-2-one derivatives to afford *trans* spiro-epoxyindoles. Azizi et al. [182] made use of ultrasound in water or polyethylene glycol to assist the rapid and sustainable catalyst-free synthesis of thiourea from the condensation between aliphatic aines and carbon disulphide. The traditional mechanical shaking method required prolonged reaction times, whereas sonicated reactions were completed within 5 min of ultrasonic irradiation. Unsymmetrical and symmetrical thiourea can be achieved with a very simple work up involving simply filtration of the precipitated thiourea or extraction with ethyl acetate for the liquid products. The reaction products were obtained in high purity and moderate to high yields (50–80%) with PEG offering a slightly higher yields and did not require further purification. This procedure has the advantaged over the traditional ones which resort to the use of the more dangerous thiophosgene and isothiocyanates.

Warfarin is one of the most effective anticoagulants used as a racemate. However, the (S)-form proved to be more active to its mirror image [183]. Warfarin can be achieved by the asymmetric Michael addition catalysed by organic primary amines and under ultrasound. The conjugate addition of the enolate from the 4-hydroycoumarin to an α,β-unsaturated ketone, with catalysis of (*S,S*)-diphenylethylenediamine afforded warfarin in 98% yield. A series of warfarin analogues were achieved in good to excellent yields (73–98%) and with good enantioselectivities (up to 76%) [184].

Acknowledgements

This work was supported by LAQV-REQUIMTE and Fundação para a Ciência e a Tecnologia through projects (PEst-C/LA0006/2013) one contract under Principal Investigator FCT (L.C. Branco) and post-doctoral grant (SFRH/BPD/111168/2015).

Author details

Luis C. Branco*, Ana M. Faisca Phillips, Maria M. Marques, Sandra Gago and Paula S. Branco

*Address all correspondence to: l.branco@fct.unl.pt

LAQV-REQUIMTE, Department of Chemistry, Faculty of Science and Technology, New University of Lisbon, Caparica Campus, Caparica, Portugal

References

[1] N. Mase, C. F. Barbas III. In water, on water, and by water: mimicking nature's aldolases with organocatalysis and water. Org. Biomol. Chem. 2010; 8:4043–4050.

[2] M. Raj, V. K. Singh. Organocatalytic reactions in water. Chem. Commun. 2009; 44:6687–6703.

[3] J. G. Hernández, E. Juaristi. Recent efforts directed to the development of more sustainable asymmetric organocatalysis. Chem. Commun. 2012; 48:5396–5409.

[4] B. H. Lipshutz, S. Ghorai. Transitioning organic synthesis from organic solvents to water. What's your E factor? Green Chem. 2014; 16:3660–3679.

[5] U. M. Lindström. Stereoselective organic reactions in water. Chem. Rev. 2002; 102:2751–2772.

[6] R. Breslow. Hydrophobic effects on simple organic reactions in water. Acc. Chem. Res. 1991; 24:159–164.

[7] S. Narayan, J. Muldoon, M. G. Finn, V. V. Fokin, H. C. Kolb, K. B. Sharpless. "On water": unique reactivity of organic compounds in aqueous suspension. Angew. Chem. Int. Ed. 2005; 44:3275–3279.

[8] B. M. Trost, C. S. Brindle. The direct catalytic asymmetric aldol reaction. Chem. Soc. Rev. 2010; 39:1600–1632.

[9] J. Mlynarski, S. Bas. Catalytic asymmetric aldol reactions in aqueous media – a 5 year update. Chem. Soc. Rev. 2014; 43:577–587.

[10] S. Bhowmick, A. Mondal, A. Ghosh, K. C. Bhowmick. Water: the most versatile and nature's friendly media in asymmetric organocatalyzed direct aldol reactions. Tetrahedron Asymmetry 2015; 26:1215–1244.

[11] B. List, R. A. Lerner, C. F. Barbas III. Proline-catalyzed direct asymmetric aldol reaction. J. Am. Chem. Soc. 2000; 122:2395–2396.

[12] A. I. Nyberg, A. Usano, P. M. Pihko. Proline-catalyzed ketone-aldehyde aldol reactions are accelerated by water. Synlett 2004; 11:1891–1896.

[13] Y. Hayashi, T. Sumiya, J. Takahashi, H. Gotoh, T. Urushima, M. Shoji. Highly diastereo- and enantioselective direct aldol reactions in water. Angew. Chem. Int. Ed. 2006; 45:958–961.

[14] N. Mase, Y. Nakai, N. Ohara, H. Yoda, K. Takabe, F. Tanaka, C. F. Barbas III. Organocatalytic direct aldol reactions in water. J. Am. Chem. Soc. 2006; 128:734–735.

[15] Y. Hayashi, S. Aratake, T. Okano, J. Takahashi, T. Sumiya, M. Shoji. Combined proline-surfactant organocatalyst for the highly diastereo- and enantioselective aqueous direct cross-aldol reaction of aldehydes. Angew. Chem. Int. Ed. 2006; 45:5527–5529.

[16] D. Gryko, W. J. Saletra. Organocatalytic asymmetric aldol reaction in the presence of water. Org. Biomol. Chem. 2007; 5:2148–2153.

[17] V. Maya, M. Raj, V. K. Singh. Highly enantioselective organocatalytic direct aldol reaction in an aqueous medium. Org. Lett. 2007; 9:2593–2595.

[18] J. F. Zhao, L. He, J. Jiang, J. Tang, L. F. Cun, L. Z. Gong. Organo-catalyzed highly diastereo- and enantio-selective direct aldol reactions in water. Tetrahedron Lett. 2008; 49:3372–3375.

[19] X. H. Chen, S. W. Luo, Z. Tang, L. F. Cun, A. Q. Mi, Y. Z. Jiang, L. Z. Gong. Organocatalyzed highly enantioselective direct aldol reactions of aldehyde with hydroxyacetone and fluoroacetone in aqueous media: the use of water to control regioselectivity. Chem. Eur. J. 2007; 13:689–701.

[20] J. Huang, X. Zhang, D. W. Armstrong. Highly efficient asymmetric direct stoichiometric aldol reactions on/in water. Angew. Chem. Int. Ed. 2007; 46:9073–9077.

[21] L. Zu, H. Xie, H. Li, J. Wang, W. Wang. Highly enantioselective aldol reactions catalyzed by a recyclable fluorous (S) pyrrolidine sulfonamide on water. Org. Lett. 2008; 10:1211–1214.

[22] Z. Tang, Z. H. Yang, L. F. Cun, L. Z. Gong, A. Q. Mi, Y. Z. Jiang. Small peptides catalyze highly enantioselective direct aldol reactions of aldehydes with

hydroxyacetone: unprecedented regiocontrol in aqueous media. Org. Lett. 2004; 24:2285–2287.

[23] F. Giacalone, M. Gruttadauria, P. L. Meo, R. Noto. New simple hydrophobic proline derivatives as highly active and stereoselective catalysts for the direct asymmetric aldol reaction in aqueous medium. Adv. Synth. Catal. 2008; 350:2747–2760.

[24] C. Zheng, Y. Wu, X. Wang, G. Zhao. Highly enantioselective organocatalyzed construction of quaternary carbon centers via cross-aldol reaction of ketones in water. Adv. Synth. Catal. 2008; 250:2690–2694.

[25] M. Markert, U. Scheffler, R. Mahrwald. Asymmetric histidine-catalyzed cross-aldol reactions of enolizble aldehydes: access to defined configured quaternary stereogenic centers. J. Am. Chem. 2009; 131:16642–16643.

[26] M. R. Vishnumaya, V. K. Singh. Highly efficient small organic molecules for enantioselective direct aldol reaction in organic and aqueous media. J. Org. Chem. 2009; 74:4289–4297.

[27] J. Li, N. Fu, X. Li, S. Luo, J. P. Cheng. Chiral primary-tertiary diamine-Brønsted acid salt catalyzed syn-selective cross-aldol reaction of aldehydes. J. Org. Chem. 2010; 75:4501–4507.

[28] G. Angelici, R. J. Corrêa, S. J. Garden, C. Tomasini. Water influences the enantioselectivity in the proline or prolinamide-catalyzed aldol addition of acetone to isatins. Tetrahedron Lett. 2009; 50:814–817.

[29] M. Raj, N. Veerasamy, V. K. Singh. Highly enantioselective synthesis of 3-cycloalkanone-3-hydroxy-2-oxindoles, potential anticonvulsants. Tetrahedron Lett. 2010; 51:2157–2159.

[30] T. C. Nugent, M. N. Umar, A. Bibi. Picolylamine as an organocatalyst template for highly diastereo- and enantioselective aqueous aldol reactions. Org. Biomol. Chem. 2010; 8:4085–4089.

[31] C. Wu, X. Fu, S. A. Li. Highly efficient, large-scale, asymmetric direct aldol reaction employing simple threonine derivatives as recoverable organocatalysts in the presence of water. Eur. J. Org. Chem. 2011; 7:1291–1299.

[32] M. De Nisco, S. Pedatella, S. Bektas, A. Nucci, R. Caputo. D-Glucosamine in a chimeric prolinamide organocatalyst for direct asymmetric aldol addition. Carb. Res. 2012; 356:273–277.

[33] Y. Qiao, Q. Chen, S. Lin, B. Ni, A. D. Headley. Organocatalytic direct asymmetric crossed-aldol reactions of acetaldehyde in aqueous media. J. Org. Chem. 2013; 78:2693–2697.

[34] A. Kumar, S. S. Chimni. Organocatalyzed direct asymmetric aldol reaction of isatins in water: low catalyst loading in command. Tetrahedron. 2013; 69:5197–5204.

[35] X. Song, A. X. Liu, S. S. Liu, W. C. Gao, M. C. Wang, J. Chang. Enantiopure azetidine-2-carboxamides as organocatalysts for direct asymmetric aldol reactions in aqueous and organic media. Tetrahedron. 2014; 70:1464–1470.

[36] S. Paladhi, J. Das, M. Samanta, J. Dash. Asymmetric aldol reaction of thiazole-carbaldehydes: regio- and stereoselective synthesis of tubuvalin analogues. Adv. Synth. Catal. 2014; 356:3370–3376.

[37] H. Y. Bae, C. E. Song. Unprecedented hydrophobic amplification in noncovalent organocatalysis "on water": hydrophobic chiral squaramide catalyzed Michael addition of malonates to nitroalkenes. ACS Catal. 2015; 5:3613–3619.

[38] X. H. Cai, B. Xie. Recent advances on organocatalysed asymmetric Mannich reactions. Arkivoc. 2013; i:264–293.

[39] I. Ibrahem, A. Córdova. Direct catalytic asymmetric anti-selective Mannich-type reactions. Chem. Commun. 2006; 16:1760–1762.

[40] L. Cheng, X. Wu, Y. Lu. Direct asymmetric three-component anti-selective Mannich reactions in a purely aqueous system. Org. Biomol. Chem. 2007; 5:1018–1020.

[41] M. Amedjkouh, M. Brandberg. Asymmetric autocatalytic Mannich reaction in the presence of water and its implication in prebiotic chemistry. Chem. Commun. 2008; 26:3043–3045.

[42] Y.-C. Teo, J.-J. Lau, M.-C. Wu. Direct asymmetric three-component Mannich reactions catalyzed by a siloxy serine organocatalyst in water. Tetrahedron Asymmetry. 2008; 19:186–190.

[43] Y. Hayashi, T. Urushima, S. Aratake, T. Okano, K. Obi. Organic solvent-free, enantio- and diastereoselective, direct Mannich reaction in the presence of water. Org. Lett. 2008; 10:21–24.

[44] T. Urushima, H. Ishikawa, Y. Hayashi. Asymmetric Mannich reaction of imines derived from aliphatic and aromatic aldehydes catalyzed by diarylprolinol silyl ether. Chem. Eur. J. 2011; 17:8273–8276.

[45] Y.-J. An, C.-C. Wang, Z.-P. Liu, J.-C. Tao. Isosteviol-proline conjugates as highly efficient amphiphilic organocatalysts for asymmetric three-component Mannich reactions in the presence of water. Helv. Chim. Acta. 2012; 95:43–51.

[46] E. Veverkova, L. Liptakova, M. Veverka, R. Šebesta. Asymmetric Mannich reactions catalyzed by proline and 4-hydroxyproline derived organocatalysts in the presence of water. Tetrahedron Asymmetry. 2013; 24:548–552.

[47] S. Moulay, A. Touati. Cycloaddition reactions in aqueous systems: a two-decade trend endeavor. C. R. Chimie. 2010; 13:1474–1511.

[48] A. B. Northrup, D. W. C. MacMillan. The first general enantioselective catalytic Diels-Alder reaction with simple α,β-unsaturated ketones. J. Am. Chem. Soc. 2002; 124:2458–2460.

[49] M. Lemay, W. W. Ogilvie. Aqueous enantioselective organocatalytic Diels-Alder reactions employing hydrazide catalysts. A new scaffold for organic acceleration. Org. Lett. 2005; 7:4141–4144.

[50] H. He, B.-J. Pei, H.-H. Chou, T. Tian, W.-H. Chan, A. W. M. Lee. Camphor sulfonyl hydrazines (CaSH) as organocatalysts in enantioselective Diels-Alder reactions. Org. Lett. 2008; 10:2421–2424.

[51] Y. Hayashi, S. Samanta, H. Gotoh, H. Ishikawa. Asymmetric Diels–Alder reactions of α,β-unsaturated aldehydes catalyzed by a diarylprolinol silyl ether salt in the presence of water. Angew. Chem. Int. Ed. 2008; 47:6634–6637.

[52] D.-Q. Xu, A.-B. Xia, S.-P. Luo, J. Tang, S. Zhang, J.-R. Jiang, Z.-Y. Xu. In situ enamine activation in aqueous salt solutions: highly efficient asymmetric organocatalytic Diels–Alder reaction of cyclohexenones with nitroolefins. Angew. Chem. Int. Ed. 2009; 48:3821–3824.

[53] Y. Ma, S. Jin, Y. Kan, Y. J. Zhang, W. Zhang. Highly active asymmetric Diels–Alder reactions catalyzed by C2-symmetric bipyrrolidines: catalyst recycling in water medium and insight into the catalytic mode. Tetrahedron 2010; 66:3849–3854.

[54] E. Neggish. Catalysis with d-block transition metals for green organic synthesis and a sustainable world. J. Synth. Org. Chem. Jpn. 2011; 69:1201.

[55] J. G. Hernandez, E. Juaristi. Recent efforts directed to the development of more sustainable asymmetric organocatalysis. Chem. Commun. 2012, 48, 5396–5409.

[56] R. Rogers, K. R. Seddon (Eds.). Ionic Liquids: Industrial Applications for Green Chemistry. American Chemical Society: Washington, 2002.

[57] J. P. Hallett, T. Welton. Room-temperature ionic liquids: solvents for synthesis and catalysis. Chem. Rev. 2011; 111:3508–3576.

[58] P. Kotrusz, I. Kmentova, B. Gotov, S. Toma, E. Solcaniova. Proline-catalysed asymmetric aldol reaction in the room temperature ionic liquid [bmim]PF_6. Chem. Commun. 2002; 21:2510–2511.

[59] T.-P. Loh, L.-C. Feng, H.-Y. Yang, J.-Y. Yang. L-Proline in an ionic liquid as an efficient and reusable catalyst for direct asymmetric aldol reactions. Tetrahedron Lett. 2002; 43:8741–8743.

[60] D. Sarkar, R. Bhattarai, A. D. Headley, B. Ni. A novel recyclable organocatalytic system for the highly asymmetric Michael addition of aldehydes to nitroolefins in water. Synthesis. 2011; 12:1993–1997.

[61] F.-F. Yong, Y.-C. Teo. Recyclabe siloxy serine organocatalyst for the direct asymmetric reactions in ionic liquids. Synth. Commun. 2011; 41:1293–1300.

[62] A. De Nino, O. Bortolini, L. Maiuolo, A. Garofalo, B. Russo, G. Sindona. A sustainable procedure for highly enantioselective organocatalyzed Diels–Alder cycloadditions in homogeneous ionic liquid/water phase. Tetrahedron Lett. 2011; 52:1415–1417.

[63] X. Ding, H.-L. Jiang, C.-J. Zhu, Y.-X. Cheng. Direct asymmetric α-amination of aldehydes with azodicarboxylates in ionic liquids catalyzed by imidazolium ion-tagged proline organocatalyst. Tetrahedron Lett. 2010; 51:6105–6107.

[64] H.-M. Guo, H.-Y. Niu, M.-X. Xue, Q.-X. Guo, L.-F. Cun, A.-Q. Mi, Y.-Z. Jiang, J.-J. Wang. L-Proline in an ionic liquid as an efficient and reusable catalyst for direct asymmetric α-aminoxylation of aldehydes and ketones. Green Chem. 2006; 8:682–684.

[65] O. V. Maltsev, A. S. Kucherenko, A. L. Chimishkyan, S. G. Zlotin. α,α-Diarylprolinol-derived chiral ionic liquids: recoverable organocatalysts for the domino reaction between α,β-enals and N-protected hydroxylamines. Tetrahedron Asymmetry. 2010; 21:2659–2670.

[66] L. Zhou, L. Wang. Chiral ionic liquid containing L-proline unit as a highly efficient and recyclable asymmetric organocatalyst for aldol reaction. Chem. Lett. 2007; 36:628–629.

[67] W. Miao, T. H. Chan. Ionic-liquid-supported organocatalyst: efficient and recyclable ionic-liquid-anchored proline for asymmetric aldol reaction. Adv. Synth. Catal. 2006; 348:1711–1718.

[68] M. L. Patil, H. Sasai. Recent developments on chiral ionic liquids: design, synthesis and applications. Chem. Rec. 2008; 8:98–108.

[69] K. Zalewska, L. C. Branco. Organocatalysis with chiral ionic liquids. Mini-Rev. Org. Chem. 2014, 11:141–153.

[70] C. Baudequin, D. Brégeon, J. Levillain, F. Guillen, J.-C. Plaquevent, A.-C. Gaumont. Chiral ionic liquids, a renewal for the chemistry of chiral solvents? Design, synthesis and applications for chiral recognition and asymmetric synthesis. Tetrahedron Asymmetry. 2005; 16:3921–3945.

[71] M. J. Earle, P. B. McCormac, K. R. Seddon. Diels–Alder reactions in ionic liquids. A safe recyclable alternative to lithium perchlorate–diethyl ether mixtures. Green Chem. 1999; 1:23–25.

[72] L. C. Branco, P. M. P. Gois, N. M. T. Lourenço, V. B. Kurteva, C. A. M. Afonso. Simple transformation of crystalline chiral natural anions to liquid medium and their use to induce chirality. Chem. Commun. 2006; 22:2371–2372.

[73] S. Luo, L. Zhang, J. P. Cheng. Functionalized chiral ionic liquids: a new type of asymmetric organocatalysts and nonclassical chiral ligands. Chem. Asian J. 2009; 4:1184–1195.

[74] M. Vasiloiu, D. Rainer, P. Gaertner, C. Reichel, C. Shroeder, K. Bica. Basic chiral ionic liquids: a novel strategy for acid-free organocatalysis. Catalysis Today 2011; 200:80–86.

[75] L. González, J. Escorihuela, B. Altava, M. I. Burguete, Santiago V. Luis. Chiral room temperature ionic liquids as enantioselective promoters for the asymmetric aldol reaction. Eur. J. Org. Chem. 2014; 24:5356–5363.

[76] A. S. Kucherenko, V. V. Gerasimchuk, V. G. Lisnyak, Y. V. Nelyubina, S. G. Zlotin. Prolinamide-derived ionic-liquid-supported organocatalyst for asymmetric mono- and bis-aldol reactions in the presence of water. Eur. J. Org. Chem. 2015; 25:5649–5654.

[77] M. S. Chauhan, P. Kumar, S. Singh. Synthesis of MacMillan catalyst tailored with ionic liquid as a recoverable catalyst for asymmetric Diels-Alder reaction. RSC Adv. 2015; 5:52636–52641.

[78] A. De Nino, L. Maiuolo, P. Merino, M. Nardi, A. Procopio, D. Roca-Lopez, B. Russo, V. Algieri. Efficient organocatalyst supported on a simple ionic liquid as a recoverable system for the asymmetric diels–alder reaction in the presence of water. ChemCatChem. 2015; 7:830–835.

[79] J. Großeheilmann, J. Bandomir, U. Kragl. Preparation of poly(ionic liquid)s-supported recyclable organocatalysts for the asymmetric nitroaldol (Henry) reaction. Chem. Eur. J. 2015; 52:18957–18960.

[80] S. Chandrasekhar, C. Narsihmulu, N. R. Reddy, S. S. Sultana. Asymmetric aldol reactions in poly (ethylene glycol) catalyzed by l-proline. Tetrahedron Lett. 2004; 45:4581–4582

[81] S. Verma, S. L. Jain, B. Sain. PEG-embedded thiourea dioxide (PEG.TUD) as a novel organocatalyst for the highly efficient synthesis of 3,4-dihydropyrimidinones. Tetrahedron Lett. 2010; 51:6897–6900.

[82] E. L. Smith, A. P. Abbott, K. S. Ryder. Deep eutectic solvents (DESs) and their applications. Chem. Rev. 2014; 114:11060–11082.

[83] Q. Zhang, K. O. Vigier, S. Royera, F. Jérôme. Deep eutectic solvents: syntheses, properties and application. Chem. Soc. Rev. 2012; 41:7108–7714.

[84] E. Massolo, S. Palmieri, M. Benaglia, V. Capriati, F. M. Perna. Stereoselective organocatalysed reactions in deep eutectic solvents: highly tunable and biorenewable reaction media for sustainable organic synthesis. Green Chem. 2016; 18:792–797.

[85] W. Leitner. Supercritical carbon dioxide as a green reaction medium for catalysis. Acc. Chem. Res. 2002; 35:746–756.

[86] I. V. Kuchurov, A. G. Nigmatov, E. V. Kryuchkova, A. A. Kostenko, A. S. Kucherenko, S. G. Zlotin. Stereodivergent Michael addition of diphenylphosphite to α-nitroalkenes in the presence of squaramide-derived tertiary amines: an enantioselective organocatalytic reaction in supercritical carbon dioxide. Green Chem. 2014; 16:1521–1526.

[87] E. V. Filatova, O. V. Turova, I. V. Kuchurov, A. A. Kostenko, A. G. Nigmatov, S. G. Zlotin. Asymmetric catalytic synthesis of functionalized tetrahydroquinolines in supercritical fluids. J. Supercritical Fluids. 2016; 109:35–42.

[88] B. Schaffner, F. Schaffner, S. P. Verevkin, A. Borner. Organic carbonates as solvents in synthesis and catalysis. Chem. Rev. 2010; 110:4554–4581.

[89] M. North, F. Pizzato, P. Villuendas. Organocatalytic, asymmetric aldol reactions with a sustainable catalyst in a green solvent. ChemSusChem. 2009; 2:862–865.

[90] M. North, P. Villuendas. A chiral solvent effect in asymmetric organocatalysis. Org. Lett. 2010; 12:2378–2381.

[91] C. Beattie, M. North, P. Villuendas. Proline-catalysed amination reactions in cyclic carbonate solvents. Molecules. 2011; 16:3420–3432.

[92] R. A. Sheldon, I. Arends, U. Hanefeld. Green Chemistry and Catalysis. Wiley-VCH, Weinheim, 2007.

[93] A. Banon-Caballero, G. Guillena, C. Najera. Solvent-free enantioselective organocatalyzed aldol reactions. Mini-Rev. Org. Chem. 2014; 11:118–128.

[94] T. P. Kumar, R. C. Shekhar, K. S. Sunder, R. Vadaparthi. Myrtanyl-prolinamide: a new chiral organocatalyst for stereoselective aldol reactions. Tetrahedron Asymmetry. 2014; 26:543–547.

[95] T. P. Kumar, N. C. Vavle, V. Patro, K. Haribabu. Phthalimido-prolinamide: a new chiral catalyst for solvent free enantioselective aldol reactions. Tetrahedron-Asymmetry. 2014; 25:457–461.

[96] E. Machuca, G. Granados, B. Hinojosa, E. Juaristi. Synthesis and evaluation of (S)-proline-containing dipeptidic organocatalysts bound to MBHA resin in asymmetric aldol reactions. Tetrahedron Lett. 2014; 56:6047–6051.

[97] E. Machuca, E. Juaristi. Organocatalytic activity of α,α-dipeptide derivatives of (S)-proline in the asymmetric aldol reaction in absence of solvent. Evidence for non-covalent π–π interactions in the transition state. Tetrahedron Lett. 2014; 56:1144–1148.

[98] J. G. Hernández, V. García-López, E. Jurasti. Solvent-free asymmetric aldol reaction organocatalyzed by (S)-proline-containing thiodipeptides under ball-milling conditions. Tetrahedron. 2012; 68:92–97.

[99] E. Machuca, Y. Rojas, E. Jurasti. Synthesis and evaluation of (S)-proline-containing α,β-dipeptides as organocatalysts in solvent-free asymmetric aldol reactions under ball-milling conditions. Asian J. Org. Chem. 2015; 4:46–53.

[100] P. Vizcaíno-Milla, J. M. Sansano, C. Nájera, B. Fiser, E. Gómez-Bengoa. Pyrimidine-derived prolinamides as recoverable bifunctional organocatalysts for enantioselective inter- and intramolecular aldol reactions under solvent-free conditions. Eur. J. Org. Chem. 2015; 12:2614–2621.

[101] M. Jörres, S. Mersmann, G. Raabe, C. Bolm. Organocatalytic solvent-free hydrogen bonding-mediated asymmetric Michael additions under ball milling conditions. Green Chem. 2013; 15:612–616.

[102] M. Jörres, I. Schiffers, I. Atodiresei, C. Bolm. Asymmetric Michael additions of α-nitrocyclohexanone to aryl nitroalkenes catalyzed by natural amino acid-derived bifunctional thioureas. Org. Lett. 2012; 14:4518–4521.

[103] M. Hestericová, R. Šebesta. Higher enantioselectivities in thiourea-catalyzed Michael additions under solvent-free conditions. Tetrahedron. 2014; 70:901–905.

[104] K. Nakashima, S.-I. Hirashima, M. Kawada, Y. Koseki, N. Tada, A. Itoh, T. Miura. Pyrrolidine-diaminomethylenemalononitrile organocatalyst for Michael additions of carbonyl compounds to nitroalkenes under solvent-free conditions. Tetrahedron Lett. 2014; 55:2703–2706.

[105] J. He, Q. Chen, B. Ni. Highly efficient asymmetric organocatalytic Michael addition of α,α-disubstituted aldehydes to nitroolefins under solvent-free conditions. Tetrahedron Lett. 2014; 55:3030–3032.

[106] A.-B. Xia, C. Wu, D.-Q. Xu, Y.-F. Wang, X.-H. Du, Z.-B. Li, Z.-Y. Xu. Prolinethiol ether catalysis in an asymmetric Michael reaction: solvent-free synthesis of functionalized monohaloalkenes. J. Org. Chem. 2013; 78:1254–1259.

[107] A. Vega-PeÇaloza, O. Sánchez-Antonio, C. G. Ávila-Ortiz, M. Escudero-Casao, E. Juaristi. An alternative synthesis of chiral (S)-proline derivatives that contain a thiohydantoin moiety and their application as organocatalysts in the asymmetric Michael addition reaction under solvent-free conditions. Asian J. Org. 2014; 3:487–496.

[108] S. Fioravanti, L. Parise, A. Pelagalli, L. Pellacani, L. Trulli. Trifluoromethyl syn- or anti-γ-amino alcohols by one-pot solvent-free Mannich-type reactions under temperature control. RSC Adv. 2015; 5:29312–29318.

[109] A. Kumar, M. K. Gupta, M. Kumar. L-Proline catalysed multicomponent synthesis of 3-amino alkylated indolesvia a Mannich-type reaction under solvent-free conditions. Green Chem. 2012; 14:290–295.

[110] M. S. Singh, S. Chowdhury. Recent developments in solvent-free multicomponent reactions: a perfect synergy for eco-compatible organic synthesis. RSC Adv. 2012; 2:4547–4592.

[111] W. Parasuk, V. Parasuk. Factors that influence stereoselectivity in proline-catalyzed Mannich reactions. Asian J. Org. Chem. 2013; 2:85–90.

[112] A. Corma, H. Garcia. Silica-bound homogenous catalysts as recoverable and reusable catalysts in organic synthesis. Adv. Synth. Cat. 2006; 348:1391–1412.

[113] S. Rostamnia, E. Doustkhah. Nanoporous silica-supported organocatalyst: a heterogeneous and green hybrid catalyst for organic transformations. RSC Adv. 2014; 4:28238–28248.

[114] A. El Kadib. Chitosan as a sustainable organocatalyst: a concise overview. ChemSusChem. 2015; 8:217–244.

[115] O. Mahé, J.-F. Brière, I. Dez. Chitosan: an upgraded polysaccharide waste for organocatalysis. Eur. J. Org. Chem. 2015; 12:2559–2578.

[116] M. Benaglia, A. Puglisi, F. Cozzi. Polymer-supported organic catalysts. Chem. Rev. 2003, 103:3401–3429.

[117] Y. Wang, X. Wang, M. Antonietti. Polymeric graphitic carbon nitride as a heterogeneous organocatalyst: from photochemistry to multipurpose catalysis to sustainable chemistry. Angew. Chem. Int. Ed. 2012; 51:68–89.

[118] B. Li, M. Chrzanowski, Y. Zhang, S. Ma. Applications of metal-organic frameworks featuring multi-functional sites. Coord. Chem. Rev. 2016; 307:106–129.

[119] D. Wang, D. Astruc. Dendritic catalysis-basic concepts and recent trends. Coord. Chem. Rev. 2013; 257:2317–2334.

[120] X. Fan, F. Zhang. Multiple roles of graphene in heterogeneous catalysis. Chem. Soc. Rev. 2015; 44:3023–3035.

[121] M. B. Gawande, P. S. Branco, R. S. Varma. Nano-magnetite (Fe_3O_4) as a support for recyclable catalysts in the development of sustainable methodologies. Chem. Soc. Rev. 2013; 42:3371–3393.

[122] R. Mrówczyński, A. Nan, J. Liebscher. Magnetic nanoparticle-supported organocatalysts – an efficient way of recycling and reuse. RSC Adv. 2014; 4:5927–5952.

[123] R. Dalpozzo. Magnetic nanoparticle supports for asymmetric catalysts. Green Chem. 2015; 17:3671–3686.

[124] M. Bartók. Advances in immobilized organocatalysts for the heterogeneous asymmetric direct aldol reactions advances in immobilized organocatalysts for the heterogeneous asymmetric direct aldol reactions. Cat. Rev. Sci. Eng. 2015; 57:192–255.

[125] R. Munirathinam, J. Huskens, W. Verboom. Supported catalysis in continuous-flow microreactors. Adv. Synth. Cat. 2015; 357:1093–1123.

[126] A. Puglisi, M. Benaglia, V. Chiroli. Stereoselective organic reactions promoted by immobilized chiral catalysts in continuous flow systems. Green Chem. 2013; 15:1790–1813.

[127] I. Atodiresei, C. Vila, M. Rueping. Asymmetric organocatalysis in continuous flow: opportunities for impacting industrial catalysis. ACS Catal. 2015; 5:1972–1985.

[128] L. A. Bivona, F. Quertinmont, H. A. Beejapur, F. Giacalone, M. Buaki-Sogo, M. Gruttadauria, C. Aprile. Thiazolium-based catalysts for the etherification of benzylic alcohols under solvent-free conditions. Adv. Synth. Cat. 2015; 357:800–810.

[129] E. J. García-Suárez, P. Moriel, C. Menéndez-Vázquez, M. A. Montes-Morán, A. B. García. Carbons supported bio-ionic liquids: stability and catalytic activity. Microporous Mesoporous Mater. 2011; 144:205–208.

[130] J. Y. Shi, C. A. Wang, Z. J. Li, Q. Wang, Y. Zhang, W. Wang. Heterogeneous organocatalysis at work: functionalization of hollow periodic mesoporous organosilica spheres with MacMillan catalyst. Chem. Eur. J. 2011; 17:6206–6213.

[131] S. M. Islam, A. S. Roy, R. C. Dey, S. Paul. Graphene based material as a base catalyst for solvent free Aldol condensation and Knoevenagel reaction at room temperature. J. Mol. Catal. A Chem. 2014; 394:66–73.

[132] Y. Zhang, Y. Zhang, L. Sun, X. Du, J. Y. Shi, W. D. Wang, W. Wang. 4-(N,N-dimethylamino)pyridine-embedded nanoporous conjugated polymer as a highly active heterogeneous organocatalyst. Chem. Eur. J. 2012; 18:6328–6334.

[133] J. Guan, B. Liu, X. Yang, J. Hu, C. Wang, Q. Kan. Immobilization of proline onto Al-SBA-15 for C–C bond-forming reactions. ACS Sustain. Chem. Eng. 2014; 2:925–933.

[134] W. Zhang, Z. Li, H. Gu, Y. Li, G. Zhang, F. Zhang, X. Fan. L-Proline Covalently Anchored on Graphene Oxide as an Effective Bifunctional Catalyst for Ketene Forming Reaction. Chem. Eng. Sci. 2015; 135:187–192.

[135] P. Wu, C. He, J. Wang, X. Peng, X. Li, Y. An, C. Duan. Photoactive chiral metal–organic frameworks for light-driven asymmetric α-alkylation of aldehydes. J. Am. Chem. Soc. 2012; 134:14991–14999.

[136] Y. Liu, X. Xi, C. Ye, T. Gong, Z. Yang, Y. Cui. Chiral metal-organic frameworks bearing free carboxylic acids for organocatalyst encapsulation. Angew. Chem. Int. Ed. 2014; 53:13821–13825.

[137] G. R. Krishnan, K. Sreekumar. First example of organocatalysis by polystyrene-supported PAMAM dendrimers: highly efficient and reusable catalyst for knoevenagel condensations. Eur. J. Org. Chem. 2008; 28:4763–4768.

[138] R. B. N. Baig, R. S. Varma. Alternative energy input: mechanochemical, microwave and ultrasound-assisted organic synthesis. Chem. Soc. Rev. 2012; 41:1559–1584.

[139] M. O. M'hamed. Ball milling for heterocyclic compounds synthesis in green chemistry: a review. Synth. Comm. 2015; 45:2511–2528.

[140] R. M. Claramunt, C. Lopez, D. Sanz, J. Elguero. Mechano heterocyclic chemistry: grinding and ball mills. In *Advances in Hetercyclic Chemistry*. San Diego, USA: Elsevier Academic Press Inc., 2014; 117–143.

[141] V. Polshettiwar, R. S. Varma. Microwave-assisted organic synthesis and transformations using benign reaction media. Acc. Chem. Res. 2008; 41:629–639.

[142] V. Polshettiwar, R. S. Varma. Aqueous microwave chemistry: a clean and green synthetic tool for rapid drug discovery. Chem. Soc. Rev. 2008; 37:1546–1557.

[143] G. Cravotto, E. Borretto, M. Oliverio, A. Procopio, A. Penoni. Organic reactions in water or biphasic aqueous systems under sonochemical conditions. A review on catalytic effects. Cat. Comm. 2015; 63:2–9.

[144] N. R. Candeias, L. C. Branco, P. M. P. Gois, C. A. M. Afonso, A. F. Trindade. More sustainable approaches for the synthesis of N-based heterocycles. Chem. Rev. 2009; 109:2703–2802.

[145] M. A. P. Martins, C. P. Frizzo, D. N. Moreira, L. Buriol, P. Machado. Solvent-free heterocyclic synthesis. Chem. Rev. 2009; 109:4140–4182.

[146] S. Toma, R. Sebesta, M. Meciarova. Organocatalytic reactions under unusual conditions. Curr. Org. Chem. 2011; 15:2257–2281.

[147] A. Bruckmann, A. Krebs, C. Bolm. Organocatalytic reactions: effects of ball milling, microwave and ultrasound irradiation. Green Chem. 2008; 10:1131–1141.

[148] A. M. Balu, D. Dallinger, D. Obermayer, J. M. Campelo, A. A. Romero, D. Carmona, F. Balas, K. Yohida, P. L. Gai, C. Vargas. Insights into the microwave-assisted preparation of supported iron oxide nanoparticles on silica-type mesoporous materials. Green Chem. 2012; 14:393–402.

[149] D. Obermayer, T. N. Glasnov, C. O. Kappe. Microwave-assisted and continuous flow multistep synthesis of 4-(pyrazol-1-yl)carboxanilides. J. Org. Chem. 2011; 76:6657–6669.

[150] T. N. Glasnov, C. Kappe. The microwave-to-flow paradigm: translating high-temperature batch microwave chemistry to scalable continuous-flow processes. Chem. Eur. J. 2011; 17:11956–11968.

[151] M. B. Gawande, V. D. B. Bonifacio, R. Luque, P. S. Branco, R. S. Varma. Solvent-free and catalysts-free chemistry: a benign pathway to sustainability. ChemSusChem. 2014; 7:24–44.

[152] A. Procopio, A. De Nino, M. Nardi, M. Oliverio, R. Paonessa, R. Pasceri. A new microwave-assisted organocatalytic solvent-free synthesis of optically enriched Michael adducts. Synlett. 2010; 12:1849–1853.

[153] M. Mojzesová, M. Mečiarová, R. Marti, R. Šebesta. Organocatalytic oxa-Diels-Alder reaction of α, β-unsaturated ketones under non-classical conditions. New J. Chem. 2015; 39: 2573–2579.

[154] A. K. Qaroush, A. S. Al-Hamayda, Y. K. Khashman, S. I. Vagin, C. Troll, B. Rieger. Highly efficient isocyanate-free microwave-assisted synthesis of [6]-oligourea. Cat. Sci. Tech. 2013; 2:2221–2226.

[155] B. Narasimhan, D. Sharma, P. Kumar. Biological importance of imidazole nucleus in the new millennium. Med. Chem. Res. 2011; 20:1119–1140.

[156] B. Jiang, X. Wang, F. Shi, S.-J. Tu, T. Ai, A. Ballew, G. Li. Microwave enabled umpolung mechanism based rapid and efficient four- and six-component formations of 2-(2 ′-azaaryl)imidazoles and anti-1,2-diarylethylbenzamides. J. Org. Chem. 2009; 74:9486–9489.

[157] X.-J. Zhang, W. Fan, L.-P. Fu, B. Jiang, S.-J. Tu. Domino [3+1+1] heterocyclization providing an efficient umpolung strategy for synthesis of 2-(2′azaaryl)imidazoles. Res. Chem. Inter. 2015; 41:773–783.

[158] G. Dessolea, R. P. Herrera, A. Ricci. H-Bonding organocatalysed Friedel-crafts alkylation of aromatic and heteroaromatic systems with nitroolefins. Synlett. 2004; 13:2374–2378.

[159] H. Kotsuki, H. Ikishima, A. Okuyama. Organocatalytic asymmetric synthesis using proline and related molecules (Part 1). Heterocycles. 2009; 75:493–529.

[160] F. Vetica, R. M. Figueiredo, M. Orsini, D. Tofani, T. Gasperi. Recent advances in organocatalytic cascade reactions toward the formation of quaternary stereocenters. Synthesis. 2015; 47:2139–2184.

[161] A. Russo, N. E. Leadbeater, A. Lattanzi. A convenient methodology for nitro-Michael addition of carbonyl compounds catalyzed by L-proline using microwave heating. Lett. Org. Chem. 2010; 7:98–102.

[162] E. M. Omara, M. B. A. R. Rahman, E. Abdulmalek, B. A. Tejo, B. Ni, A. D. Headley. Optimization of microwave-assisted Michael addition reaction catalyzed by L-proline

in ionic liquid medium using response surface methodology. Synth. Commun. 2014; 44:381–398.

[163] D. Enders, R. Krull, W. Bettray. Microwave-assisted organocatalytic quadruple domino reactions of acetaldehyde and nitroalkenes. Synthesis. 2010; 4:567–572.

[164] T. Baumann, M. Baechle, C. Hartmann, S. Braese. Thermal effects in the organocatalytic asymmetric alpha-amination of disubstituted aldehydes with azodicarboxylates: a high-temperature organocatalysis. Eur. J. Org. Chem. 2008; 13:2207–2212.

[165] J. Alemán, A. Parra, H. Jiang, K. A. Jørgensen. Squaramides: bridging from molecular recognition to bifunctional organocatalysis. Chem. Eur. J. 2011; 17:6890–6899.

[166] M. Sanchez-Rosello, C. Mulet, M. Guerola, C. del Pozo, S. Fustero. Microwave-assisted tandem organocatalytic peptide-coupling intramolecular aza-Michael reaction: alpha, beta-unsaturated N-acyl pyrazoles as Michael acceptors. Chem. Eur. J. 2014; 20:15697–15701.

[167] T. Naicker, P. I. Arvidsson, H. G. Kruger, G. E. M. Maguire, T. Govender. Microwave-assisted synthesis of guanidine organocatalysts bearing a tetrahydroisoquinoline framework and their evaluation in Michael addition reactions. Eur. J. Org. Chem. 2012; 17:3331–3337.

[168] T. Boddaert, Y. Coquerel, J. Rodriguez. Organocatalytic activity of N-heterocyclic carbenes in the Michael addition of 1,3-dicarbonyl compounds: application to a stereoselective spirocyclization sequence. Adv. Synth. Cat. 2009; 351:1744–1748.

[169] O. Bortolini, G. Fantin, M. Fogagnolo, P. P. Giovannini, A. Massi, S. Pacifico. Thiazolium-catalyzed intermolecular Stetter reaction of linear and cyclic alkyl alpha-diketones. Org. Biomol. Chem. 2011; 9:8437–8444.

[170] F. X. Liao, Y. G. Wang, Q. Zhu. Catalytic microwave-promoted direct aldol condensation using resin-bound secondary amine. Synth. Commun. 2014; 44:161–169.

[171] N. Gangwar, V. K. Kasana. 3,4-Dihydropyrimidin-2(1H)-one derivatives: organocatalysed microwave assisted synthesis and evaluation of their antioxidant activity. Med. Chem. Res. 2012; 21:4506–4511.

[172] G. Zhang, N. Ma, B. Jiang, F. Shi, S.-J. Tu. Thiosalicylic acid catalyzed multicomponent reactions: microwave-assisted synthesis of new extended angular fused azaheterocycles. Synthesis. 2010; 23:3993–3998.

[173] N. Seus, L. C. Goncalves, A. M. Deobald, L. Savegnago, D. Alves, M. W. Paixao. Synthesis of arylselanyl-1H-1,2,3-triazole-4-carboxylates by organocatalytic cycloaddition of azidophenyl arylselenides with beta-keto-esters. Tetrahedron. 2012; 68:10456–10463.

[174] D. Tejedor, L. Cotos, D. Marquez-Arce, M. Odriozola-Gimeno, M. Torrent-Sucarrat, F. P. Cossio, F. Garcia-Tellado. Microwave-assisted organocatalyzed rearrangement of propargyl vinyl ethers to salicylaldehyde derivatives: an experimental and theoretical study. Chem. Eur. J. 2015; 21:18280–18289.

[175] M. Hoffmann, S. Deshmukh, T. Werner. Scope and limitation of the microwave-assisted catalytic wittig reaction. Eur. J. Org. Chem. 2015; 20:4532–4543.

[176] T. Werner, M. Hoffmann, S. Deshmukh. First enantioselective catalytic wittig reaction. Eur. J. Org. Chem. 2014; 30:6630–6633.

[177] T. Werner, M. Hoffmann, S. Deshmukh. First microwave-assisted catalytic wittig reaction. Eur. J. Org. Chem. 2014; 30:6873–6876.

[178] J. McNulty, D. McLeod, P. Das, C. Zepeda-Velazquez. Wittig reactions of trialkylphosphine-derived ylides: new directions and applications in organic synthesis. Phosphorus Sulfur Silicon Related Elements. 2015; 190:619–632.

[179] D. M. Jackson, R. L. Ashley, C. B. Brownfield, D. R. Morrison, R. W. Morrison. Rapid conventional and microwave-assisted decarboxylation of L-histidine and other amino acids via organocatalysis with R-carvone under superheated conditions. Synth. Commun. 2015; 45:2691–2700.

[180] T. J. Mason. Ultrasound in synthetic organic chemistry. Chem. Soc. Rev. 1997; 26:443–451.

[181] C. Mangilal, P. Anang, S. Ratnesh, A. N. Vipin. Quinine as an organocatalytic dual activator for the diastereoselective synthesis of spiro-epoxyoxindoles. Tetrahedron Lett. 2013; 54:7119–7123.

[182] N. Azizi, A. Rahimzadeh-Oskooee, Z. Yadollahy, A. G. Ourimi. Ultrasound-assisted rapid sustainable synthesis of substituted thiourea. Monatsh. Chem. 2014; 145:1675–1680.

[183] D. Fishelovitch, C. Hazan, S. Shaik, H. J. Wolfson, R. Nussinov. Structural dynamics of the cooperative binding of organic molecules in the human cytochrome P4503A4. J. Am. Chem. Soc. 2007; 129:1602–1611.

[184] M. Rogozinska-Szymczak, J. Mlynarski. Asymmetric synthesis of warfarin and its analogues on water. Tetrahedron Asymmetry. 2014; 25:813–820.